THE INTENSITY INTERFEROMETER

its Application to Astronomy

THE INTENSITY INTERFEROMETER

its Application to Astronomy

R. HANBURY BROWN, F.R.S.

Professor of Physics (Astronomy)
University of Sydney

TAYLOR & FRANCIS LTD
LONDON

HALSTED PRESS
division
John Wiley & Sons Inc.
NEW YORK—TORONTO

1974

First published 1974 by Taylor & Francis Ltd
London and Halsted Press (a division of
John Wiley & Sons Inc) New York

Taylor & Francis   ISBN   0   85066 072 6

Printed and bound in Great Britain by
Taylor & Francis Ltd
10–14 Macklin Street, London WC2B 5NF

Library of Congress Cataloging in Publication Data
Brown, Robert Hanbury, 1916–
    The intensity interferometer.
      1. Interferometer. 2. Astronomy. I. Title.
QB86.B73 1974   522'.5   74–14878
ISBN 0–470–10797–9

Preface

In this book I have tried to write a history of the intensity interferometer together with a brief account of the theory, practice and application of this new instrument to a classical problem in astronomy—the measurement of the apparent angular diameters of stars. Much of the scientific material has already appeared in print but it is scattered through more than twenty years of scientific literature. My aim has been to collect it into one book and to tell how and why the work came to be done.

There are at least three good reasons for doing this. Firstly, the intensity interferometer has proved to be of lively interest to physicists and incomprehensible to astronomers; there should be a single written account appropriate to both groups. Secondly, the measurements of the angular diameters of stars made at Narrabri Observatory have been widely used by astrophysicists and there is no other existing instrument capable of checking them. It is therefore doubly important that there should be a detailed account of exactly how the observations were made and analysed so that their credibility can be assessed. Finally, I want to preserve our experience so that other groups, perhaps in the Northern Hemisphere, will one day follow our lead. Those of us who have tried to build on the experience gained with the unsuccessful 50 ft. stellar interferometer built at Mount Wilson by Hale and Pease have found that although there are descriptions of the instrument, like so many scientific papers, they don't tell us all we need to know—what checks were made and what went wrong. With this in mind I hope the reader will excuse some of the details in this book which, at first sight, may seem unnecessary.

During my life I have been fortunate to take part in two other scientific adventures: the early work on radar and the beginnings of radio-astronomy. Books about those pioneering days have always seemed to me to leave out the essential ingredient of such adventures—the sense of purpose and excitement which transfigures the hardest work and dullest routine. Telling my own story has made me well aware that a lifetime of writing scientific papers is not a good training for conveying that same sense of adventure which was felt by all who helped to build the stellar interferometer at Narrabri. If only for one moment I could take you into the great plains of northern New South Wales and show you this elegant instrument looking at the stars, I believe you would understand what is missing from my book. Not only was the work we did at Narrabri worth while in strictly scientific terms, but in personal terms it was exciting, exhausting and profoundly rewarding.

The success of a project lasting more than two decades has involved the help and support of many people over the years. Much of the work described in this book has been done by my colleagues, Dr R. Q. Twiss, Dr J. Davis and Dr L. R. Allen.

Others who have "done time" at Narrabri are Dr D. Herbison-Evans and Dr C. Hazard and a small galaxy of postgraduate students: Dr M. J. Yerbury, Mr J. M. Rome, Mr R. J. W. Lake, Mr D. W. Keenan, Dr R. J. Webb and Mr R. K. Outhred. Mr Gifford has taken care of the maintenance of the Observatory for ten years, Mr S. Owens has looked after the electronics and Miss V. Raymond has been throughout, secretary to the Department of Astronomy in the University of Sydney.

The original design of the instrument owes much to the staff of Dunford and Elliott (Sheffield), Mullard Ltd. (Salford) and MacDonald, Wagner and Priddle (North Sydney). The financial support was provided by the Department of Scientific and Industrial Research (now the Science Research Council), the Nuclear Research Foundation (now the Foundation for Research in Physics) within the University of Sydney, the Office of Scientific Research of the United States Air Force, the Australian Research Grants Committee and the Research Grants Committee of the University of Sydney.

I am grateful to Sir Bernard Lovell for giving the original work at Jodrell Bank his support and to Professor Harry Messel who, as Head of the School of Physics in the University of Sydney, has always supported us with characteristic enthusiasm. Among our many friends overseas, I must mention

Dr D. M. Popper of U.C.L.A., Dr E. P. Ney of the University of Michigan and Dr D. C. Morton of Princeton, all of whom have worked at Narrabri.

Finally I thank my wife Heather for making our life in the bush so happy and for her help with everything described in this book except the mathematics.

Narrabri Observatory   R. HANBURY BROWN
New South Wales
Australia

June 1973

List of Symbols

A	area of light detector or mirror
$a(\nu)$	amplitude of Fourier component of light wave
a	fraction of photomultiplier anode current due to stray light and dark current
B	apparent blue magnitude of star
B_0	bandwidth of light (eqn. (5.2))
b_V	electrical bandwidth (eqn. (5.10))
c	velocity of light
$c(d)$	correlation observed in 100 s at a baseline d
$c_0(d)$	observed correlation corrected for zero-drift (eqn. (10.1))
$c_N(d)$	normalized correlation (eqn. (10.4))
C_N	zero-baseline correlation normalized to value expected for a single unresolved star
D	zero-drift of correlator
d	baseline length
E	amplitude of electric vector of light wave
e	charge on the electron
e_1, e_2	voltages
$F(f)$	voltage gain at frequency f
F_λ	Absolute monochromatic flux emitted per unit area of stellar surface at wavelength λ
f	frequency of electric current or voltage
f_λ	absolute monochromatic flux received from a star at wavelength λ
G_c	electrical gain of correlator
$G(\nu)$	mutual spectral density of two light beams (eqn. (3.12))
$g(\nu)$	transmittance of optical system at frequency ν
h	the Planck constant
I	light intensity
$I(S)$	light intensity per unit area of source
i (§ 4)	angle of inclination of orbital plane of a binary star
i_1, i_2	electric currents

j_c	wave noise (eqn. (4.18))
j_n	shot noise (eqn. (4.19))
k	the Boltzmann constant
l_0	coherence length of light (eqn. (3.33))
N, n	number of counts of pulses
$N(T)$	r.m.s. noise or uncertainty after a time T
$n(\nu)$	number of photons per unit time, per unit area, per unit light bandwidth at frequency ν
R	radius of star
R_1, R_2	distances from observer to points on a light source
(S/N)	signal to noise ratio
T, t	time
T_e	effective temperature of a star
u	limb-darkening coefficient (eqn. (10.12))
V	apparent visual magnitude of star
$V(d)$	fringe visibility with baseline d
$V(t)$	electric vector of light wave
$W(f)$	spectral density of electrical fluctuations
z	(§5.6) zenith angle, (§5.8) coordinate parallel to line from observer to source
α	quantum efficiency
β	ratio of brightness of components of binary star
β_0	polarization factor (eqn. (5.7))
$\Gamma^2(d)$	normalized correlation factor (eqn. (5.8))
$\left.\begin{array}{c}\Gamma(\tau) \\ \Gamma(d)\end{array}\right\}$	mutual coherence function (eqn. (3.16))
$\gamma_d, \gamma_{12}(\tau)$	complex degree of coherence (eqn. (3.18))
Δ	partial coherence factor (eqn. (5.11))
Δm	difference in brightness of components of a binary star
$\Delta\nu$	light bandwidth
Δf	electrical bandwidth
ΔI	intensity fluctuation

δ	excess noise in correlator (eqn. (5.15))
ϵ	fraction of correlation lost in correlator
η	spectral density of cross-correlation frequency response (eqn. (5.16))
Θ_0	black-body temperature of source
θ_e	reciprocal effective temperature of star
θ_{UD}	angular diameter of star, equivalent uniform disc
θ_{LD}	true angular diameter of star (corrected for limb-darkening)
θ_s	angular separation between components of a binary star
λ	wavelength of light
μ	§5.4, 5.5, 5.6 gain of first dynode of photomultiplier; §5.8 refractive index
ν	frequency of light wave
$\rho(\nu)$	atmospheric transmission at frequency ν
σ_{OBS}	r.m.s. uncertainty (eqn. (9.1)) in 100 s observation of correlation
σ_{STD}	standard uncertainty (eqn. (9.2)) in correlation
τ	time delay
τ_c	resolving time
τ_0	coherence time of light (eqn. (3.33))
ϕ	phase-shift
ω	angular frequency
Ω	solid angle

Contents

Preface   v

List of Symbols   ix

Chapter 1   THE STORY OF HOW AND WHY THE
              STELLAR INTENSITY INTERFEROMETER
              AT NARRABRI CAME TO BE BUILT

1.1   The Historical Problem   1
1.2   The First Intensity Interferometer   2
1.3   The Theory of an Intensity Interferometer for Light
      Waves   4
1.4   Opposition to the Theory   5
1.5   A Pilot Model for an Optical Stellar Interferometer   9
1.6   Two more Laboratory Experiments   10
1.7   Financing and Building the Narrabri Stellar
      Interferometer   10
1.8   Installation of the Interferometer at Narrabri   14
1.9   Preliminary Test, Pilot Programme and Teething
      Troubles   16
1.10 The Main Work of the Observatory   18
1.11 The Future   21

Chapter 2   A SIMPLE EXPLANATION OF HOW AN
              INTENSITY INTERFEROMETER WORKS   22

2.1   Introduction   22
2.2   Michelson's Stellar Interferometer   22
2.3   The Intensity Interferometer   25

Chapter 3   THE THEORY OF COHERENT LIGHT   32

3.1   A Mathematical Description of Light   32
3.2   The Interference of Two Partially Coherent Beams
      of Light   33
3.3   Spatial Coherence—The Dependence of Coherence
      on the Angular Size of the Source   35
3.4   Temporal Coherence—The Dependence of Coherence
      on Path Difference   37
3.5   The Correlation between Fluctuations of Intensity   39

Chapter 4   THE PRINCIPLES OF THREE TYPES
OF INTERFEROMETER   41
4.1   The Principles of Michelson's Interferometer   41
4.1.1 Fringe visibility   41
4.1.2 The effects of path differences   43
4.2   The Principles of an Intensity Interferometer using a
Linear Multiplier   45
4.2.1 The fluctuations in the output of a photoelectric
detector   45
4.2.2 The correlation between the fluctuations in the
output currents of two photoelectric detectors   48
4.2.3 The signal-to-noise ratio   50
4.2.4 The effects of path difference or time delay   51
4.3   The Principles of an Intensity Interferometer using
a Coincidence Counter   54
4.3.1 Fluctuations in the output of a photon-counting
detector   54
4.3.2 Coincidence-counting   56
4.3.3 The signal-to-noise ratio   58

Chapter 5   THE THEORY OF A PRACTICAL
STELLAR INTENSITY INTERFEROMETER   59
5.1   Introduction   59
5.2   Correlation between Small Apertures   59
5.3   Correlation between Large Apertures   60
5.4   The Noise   63
5.5   The Signal-to-Noise Ratio   63
5.6   The Maximum Possible Signal-to-Noise Ratio   63
5.7   A Theoretical Estimate of the Effects of Čerenkov Light   64
5.8   A Brief Theory of the Effects of Atmospheric
Scintillation   67
5.8.1 A mathematical model of scintillation   67
5.8.2 Phase-scintillation   69
5.8.3 Amplitude scintillation   70
5.8.4 Angular scintillation   71
5.8.5 Summary of the effects of scintillation   72

xiii

Chapter 6	LABORATORY TESTS	73
6.1	Tests of a Linear Multiplier Intensity Interferometer	73
6.1.1	The optical system	73
6.1.2	The correlator	74
6.1.3	Experimental procedure and results	77
6.1.4	Comparison between theory and experiment	78
6.2	Tests of a Coincidence-counting Intensity Interferometer	79
6.2.1	Introduction	79
6.2.2	The experiment of Twiss and Little	80
6.2.3	Other experiments	83
Chapter 7	TWO EARLY INTENSITY INTERFEROMETERS	84
7.1	A Radio Intensity Interferometer	84
7.2	An Optical Stellar Intensity Interferometer	88
7.2.1	The equipment	88
7.2.2	Experimental procedure and results	90
7.2.3	Comparison between Theory and Observation	91
7.2.4	Discussion and conclusion	93
Chapter 8	THE NARRABRI STELLAR INTERFEROMETER	94
8.1	General Layout	94
8.2	The Reflectors	94
8.3	Guiding and Control	100
8.4	The Phototubes	102
8.5	The Correlator	103
8.6	An Alternative Design of Correlator	109
Chapter 9	HOW THE OBSERVATIONS AT NARRABRI WERE MADE	111
9.1	Focusing the Reflectors	111
9.2	Equalizing the Time Delays	112
9.3	Measuring the Zero-drift, Gain and Noise Level of the Correlator	112
9.4	Choosing the Baselines and Exposure Times	114
9.5	Observational Procedure	115

Chapter 10   ANALYSING THE DATA   118
    10.1   The Normalized Correlation   118
    10.2   Estimating the Uncertainty in the Normalized
        Correlation   120
    10.3   Finding the Angular Diameter of Single Stars   122
  10.3.1   The angular diameter of the equivalent uniform disc   122
  10.3.2   The effects of limb-darkening   123
  10.3.3   The effects of multiple stars   126
  10.3.4   The effects of stellar rotation   128
  10.3.5   The effects of polarization   129

Chapter 11   RESULTS   132
    11.1   Angular Diameters and Zero-baseline Correlations   132
    11.2   Radii, Fluxes and Temperatures   133
  11.2.1   True angular diameters   133
  11.2.2   Radii   133
  11.2.3   Emergent fluxes   136
  11.2.4   Effective temperatures   137
    11.3   The Detection of Multiple Stars   139
    11.4   The Spectroscopic Binary: Spica (α Vir)   141
  11.4.1   Introduction   141
  11.4.2   Method of observation   142
  11.4.3   Analysis of observations   142
  11.4.4   Results   144
  11.4.5   Discussion of the results on Spica (α Vir)   146
    11.5   The Emission-line Star γ Velorum   148
    11.6   Limb-darkening of Sirius   149
    11.7   The Rotation of Altair   151
    11.8   Polarization Tests on Rigel   154
    11.9   Observations of Čerenkov Light Pulses   155
  11.10   A Search for Sources of False Correlation   157
  11.11   The Effects of Atmospheric Scintillation   159
  11.12   Signal-to-Noise Ratios   160

Chapter 12   FUTURE POSSIBILITIES   162
    12.1   Introduction   162
    12.2   The Measurement of Emergent Fluxes, Effective
            Temperatures and Radii of Single Stars   163
    12.3   Double Stars   164
    12.4   Cepheid Variables   165
    12.5   Emission-line Stars and Rotating Stars   166
    12.6   Limb-darkening, Polarization and Extended
            Atmospheres   167
    12.7   Interstellar Extinction   167
    12.8   The Specifications of a New Instrument   168
    12.9   A Possible Configuration of a New Interferometer   169
    12.10 Summary   171
           References   173
           Abbreviations of constellations   177
           Chart showing constellations and stars   178
           Names of stars   179

Index   180

CHAPTER 1

the story of how and why the stellar intensity interferometer at Narrabri came to be built

1.1 *The Historical Problem*

In essence, this story is about the discovery and development of a new technique of physical measurement, *intensity interferometry*, and its application to the specific problem of measuring the extremely small angles subtended at the Earth by bright stars.   Although the technique emerged from work in the comparatively new field of radio-astronomy the problem is an old one in the history of astronomy which has proved peculiarly difficult to solve—in fact it is not completely solved yet. There have been many attempts in the past to find the apparent angular diameter of the stars and it is worth while to recall three which illustrate some of the difficulties that the problem holds.

The first was an experimental attack made, characteristically, by Galileo.   He suspended a fine cord vertically and measured the distance at which he had to stand from the cord so that it just occulted the image of the first magnitude star Vega.   A brief account of the experiment is given in his *Dialogue concerning the Two Chief World Systems* and it is interesting to read how carefully it was done; he even allowed for the convergence of light in his eye.   Galileo reached the conclusion that the angular diameter of Vega is 5 seconds of arc, which was a significant advance on the currently accepted value of about 2 minutes of arc and he used this measurement to considerable effect in his trenchant criticism of the objections to the Copernican Theory.   When the angular diameter of Vega was measured again (about 350 years later at Narrabri), it was found to be about 3×10^{-3} seconds of arc; Galileo's result was roughly 1500 times too large and one wonders why.   At first sight, his experiment looks very simple to do but when you work out the actual details—the distance from the cord, the angular motion of Vega, etc.—you find that the observation must have been carried out skilfully to get a result even as small as 5 seconds of arc.   The actual value measured by Galileo was, presumably, not a measurement of the size of Vega but a measurement of the local angular scintillation due to the atmosphere.

A theoretical attempt to estimate the angular size of a first magnitude star was made by Newton.   He argued that if one is prepared to assume that the Sun is a body similar to the fixed stars (an assumption which is still surprising to many people) and if it were to be removed to a distance

1

at which it appears as a first magnitude star, then its angular diameter would be about 2×10^{-3} seconds of arc. This estimate is about two-thirds of the value we have now established for Vega, so we must regard Newton's effort to solve the problem as approaching much closer to the right answer.

The third attempt of importance was the first successful measurement of a star. It was made by Michelson and Pease on 13 December 1920, using the 20 ft stellar interferometer at Mount Wilson; they found the angular diameter of the supergiant star Betelgeuse to be 47×10^{-3} seconds of arc. Although six stars, all cool giants or supergiants, were measured with that instrument, all subsequent efforts to extend the work to other stars have failed. For example, in 1930, F. G. Pease built a larger version of Michelson's interferometer with a 50 ft beam but could not make it give reliable results and the work was abandoned in 1937.

This slow progress in solving a classical problem is not surprising when you appreciate the technical difficulties involved. To begin with, the angles subtended by stars at the Earth are extremely small; for example, to find the angular size of even a few of the hot stars one must measure angles of the order of 10^{-4} seconds of arc. This means that any instrument that one uses must be capable of working with baselines several hundred metres long. A second complication is that turbulence in the Earth's atmosphere distorts the light waves reaching us from the stars to such an extent that, even when viewed in a large telescope, the image of a star is blurred into a shapeless patch which is enormous compared with the true angular size of the star and is seldom less than 1 second of arc across.

These two major difficulties have for many years arrested the development of Michelson's interferometer and stood in the way of further progress towards finding the angular size of the stars.

1.2 The First Intensity Interferometer

Both problems—the need for long baselines and the effects of atmospheric turbulence—were solved by the invention or, if you prefer the word, the discovery of intensity interferometry. Quite simply, I thought of the idea late one night in 1949. I was trying to design a radio interferometer which would solve the intriguing problem of measuring the angular sizes of the two most prominent radio sources in the sky, Cygnus A and Cassiopeia A. At that time we knew only that their angular sizes could not be much greater than a few minutes of arc but we had no evidence as to how small they might be. If, as some people thought, they proved to be as small as the visible stars, then at metre wavelengths we should need two stations at the ends of the Earth. Was it feasible to build a radio interferometer with a baseline that could be extended if necessary from tens to hundreds, or perhaps thousands, of kilometres? The immediate technical difficulty

2

in adapting conventional designs was to provide a coherent oscillator at the two distant points, and I started to wonder if this was really necessary. Could one perhaps compare the radio waves received at two points by some other means? As an example, I imagined a simple detector which demodulated waves from the source and displayed them as the usual noise which one sees on a cathode-ray oscilloscope. If one could take simultaneous photographs of the noise at two stations, would the two pictures look the same? This question led directly to the idea of the correlation of intensity fluctuations and to the principle of intensity interferometry.

A few days later, having convinced myself that the idea, unlike most late-night discoveries, was sound, I sought the help of Richard Twiss to put the whole scheme on a formal mathematical basis. After some characteristic pooh-poohing based, I remember, on a minor mistake in one of his integrals, Richard seized on the idea with alacrity and produced, on sheet after sheet of paper, a rigorous and quantitative theory of the radio intensity interferometer which he sent to me in instalments through the post (Hanbury Brown and Twiss, 1954). Before long we were ready to design a practical system.

Our first step was to join with R. C. Jennison in building at Jodrell Bank a radio intensity interferometer at a frequency of 125 MHz and to test it by measuring the angular diameter of the quiet Sun. After some preliminary anxieties, due to the fact that the two halves of the antenna were connected the wrong way round, radiation was received from the Sun as it passed through the aerial beam and much to our relief the predicted correlation was recorded. The scheme worked exactly as we had hoped.

The next step, again with R. C. Jennison joined by M. K. Das Gupta, was to apply this interferometer to the measurement of the two radio sources in Cygnus and Cassiopeia. Since we had no idea how long the baseline should be, we limited the bandwidth of the correlated post-detector noise to only 2000 Hz. We proposed to transmit this noise from one station to the other by telephone line or radio link. If very long baselines proved to be necessary, then the bandwidth would have to be even narrower, the signals being recorded at the two stations separately; the records would have to be synchronized by some radio signal and subsequently brought together and correlated in a later operation. We decided to start by transmitting the correlated noise as a modulation on a radio link; this seemed to be more promising than battling with the rural telephone network in that part of Cheshire. In 1952 the first observations of both sources were made with the two stations of the interferometer close together and the theoretical correlation was observed. Then one station, complete with 500 square metres of antenna, was loaded on to a lorry, and set up in a farmyard about a mile from Jodrell Bank. Again all went well, and in subsequent measurements the remote station worked its way across Cheshire, farm by farm. To our great satisfaction the whole thing worked perfectly and the measurements were steady and repeatable; but to our disappointment it was all over much too quickly. Both Cassiopeia A and Cygnus A proved to be so large that they were resolved with baselines of only a few kilometres, making our elaborate preparations for very long baselines unnecessary; we had used a sledge-hammer to crack a nut. We could have done the whole job by making minor alterations to a conventional interferometer, which would have been comparatively easy to develop and would have required a much smaller mobile antenna array. Nevertheless, the actual measurements on the two radio sources were reliable and, at that time, new and important (Hanbury Brown, Jennison and Das Gupta, 1952); they were confirmed by measurements with more conventional interferometers in Cambridge and Sydney and they have stood the test of time.

At the beginning of this programme we had thought that the sole advantage of an intensity interferometer, compared with the radio version of Michelson's interferometer, was that it did not require mutually coherent local oscillators at the separated stations and was therefore peculiarly suitable for extremely long baselines. (At that time the highly stable oscillators with rubidium or caesium, which are now used for long baselines, were not invented.) However, as we watched our interferometer at work, we noticed that when the radio sources were scintillating violently, due to ionospheric irregularities, the measurements of correlation were not significantly affected. Richard Twiss investigated the theory of this surprising effect and confirmed that it was to be expected. We had overlooked one of the principal features of an intensity interferometer—the fact that it can be made to work through a turbulent medium.

It was this last result which prompted us to enquire whether an intensity interferometer could be made to work at the wavelength of light. If we could make an optical interferometer with the very long baselines necessary to resolve main-sequence stars *and* with the ability to work reliably through the Earth's atmosphere, then we could overcome the two main obstacles which had prevented the development of Michelson's work. We decided to look into the theory of an intensity interferometer for light waves.

1.3 *The Theory of an Intensity Interferometer for Light Waves*

In principle, the theory is the same for all wavelengths; however, in terms of practical formulae the domains of radio and light are quite distinct. Radio engineers, before the advent of masers, thought of radio waves as waves and not as a shower of photons. A way of picturing this difference is to say that because the energy of the radio photon is so small and there are so many photons, the energy comes smoothly and not in bursts. The fluctuations in the output of a simple square-law detector, exposed to these waves through an antenna, can therefore be calculated in terms of classical waves. We say that the fluctuations in its output are principally due to 'wave noise' and not to 'photon noise'. By contrast, at optical wavelengths, the energy of the individual photon is much greater and there are relatively few photons, so that we can no longer neglect the fact that the energy comes in bursts. In this case the situation is reversed; the fluctuations in the output of the detector are due principally to 'photon noise' and not 'wave noise'. This matter is expressed more formally in chapter 4.

After a good deal of argument the formulae for an optical interfero-meter were produced by Richard Twiss. Disappointingly, they seemed to show that an optical intensity interferometer would be absurdly expensive. To measure a first magnitude star we should need two telescopes at least 2·5 m in diameter and to make matters worse, one of these must be mobile. Clearly such a project was impracticable and reluctantly we let the matter drop.

It took us six months to realize that although we should certainly need two very large telescopes, they could be extremely crude by astronomical standards. Their function would simply be to collect the light from the star like rain in a bucket and pour it on to the detector; there was no need to form a conventional image. In practice this meant that the whole problem was transformed into one of reasonable cost since our telescopes need only be like the paraboloids used for radio-astronomy, but with light-reflecting surfaces. The necessary precision of these surfaces would be governed by the maximum permissible field of view and not, as at radio wavelengths, by the beamwidth; for bright stars a field of view of several minutes of arc would be tolerable and this could be achieved with the sort of structures which were used by radio-astronomers for microwaves.

With renewed enthusiasm we returned to establish the detailed theory of an optical interferometer, and immediately ran into a barrage of criticism.

1.4 *Opposition to the Theory*

Our original theory was clearly correct at radio wavelengths but when it came to light waves there were one or two lingering doubts in our own minds and several firmly entrenched doubts in the minds of others. The trouble of course was due to worrying about photons. As I have already pointed out, radio engineers in those days looked on radio waves as being simply waves and our theory of the radio intensity interferometer was accepted without question. But when we came to deal with physicists, all sorts of queries were raised. One group of objections was concerned with the validity of our semi-classical model of photoelectric emission. We had assumed that the probability of emission of a photoelectron is proportional to the instantaneous square of the electric vector of the incident light wave treated classically.

Further, we had assumed that there is no significant delay in the photoelectric process and that in the output current all the components of the envelope of light, at least up to 100 MHz, would be present with their correct phases and amplitudes. At that time there was no sufficiently detailed quantum-mechanical treatment of photoelectric emission and we justified our semi-classical picture on general theoretical grounds and by appeal to an experiment performed by Forrester, Gudmundsen and Johnson (1955). In their experiment, an isotope lamp was placed in a magnetic field and the light illuminated a phototube. The frequency components of the light, due to Zeeman splitting, beat with each other in the photoelectric detector and their difference frequency ($\sim 10^{10}$ Hz) was detected in the photocurrent. This experiment demonstrated the formation of very high beat-frequencies in photoelectric detection and indicated that any delays in the process are less than 10^{-10} s.

One particular doubt about the reliability of our model is worth recalling because the solution proved to be instructive. It was pointed out by Fellgett (1957) and also by Clark Jones, in a private letter, that our semi-classical analysis gave an apparently incorrect result for the fluctuations in the temperature of a grey body in thermal equilibrium with an isothermal enclosure. The established formula, based on thermodynamics, gave the fluctuations Δm in the number of photons m exchanged by the body in each small frequency range as

$$\overline{\Delta m^2} = m[1 + n/N] \qquad (1.1)$$

where n is the mean density of photons in the enclosure and N is the mean density of Bose cells. The first term in the brackets corresponds to the fluctuations in a classical assembly of particles, and the second term to the 'wave noise'. Applied to the same problem our analysis gave

$$\overline{\Delta m^2} = m[1 + \epsilon n/N] \qquad (1.2)$$

where ϵ is the emissivity of the body. It therefore agreed with the established formula for the case of a black body ($\epsilon = 1$), but gave a lower 'wave noise' for a grey body. This problem worried us for some time; we felt certain that the 'thermodynamic' formula was wrong but we could not see why. In a first unsuccessful attempt to answer the objection we published some rather misguided criticisms of the thermodynamic argument (Hanbury Brown and Twiss, 1957 a). The question was finally cleared up, largely due to Richard Twiss who published a joint note with Fellgett and Clark Jones (Fellgett, Clark Jones and Twiss, 1959), giving an explanation of the discrepancy. Briefly, the 'thermodynamic' argument assumes that the total fluctuations in the temperature of the body can be treated as the sum of two independent streams of radiation, one absorbed and one emitted. This is incorrect because the incident, emitted, and reflected streams of radiation interact; when this interaction is taken into account the formulae can be reconciled.

Another stream of objections about photons were both instructive and entertaining. Our whole argument was based on the idea that the fluctuations in the outputs of two photoelectric detectors must be correlated when they are exposed to a plane wave of light. We had shown that this must be so by a semi-classical analysis in which light is treated as a classical wave and in this picture there is no need to worry about photons—the quantization is introduced by the discrete energy levels in the detector. However, if one must think of light in terms of photons then, if the two pictures are to give the same result, one must accept that the times of arrival of these photons at the two separated detectors are correlated—they tend to arrive in pairs. Now, to a surprising number of people, this idea seemed not only heretical but patently absurd and they told us so in person, by letter, in publications, and by actually doing experiments which claimed to show that we were wrong. At the most basic level they asked how, if photons are emitted at random in a thermal source, can they appear in pairs at two detectors? At a more sophisticated level the enraged physicist would brandish some sacred text, usually by Heitler, and point out that the number n of quanta in a beam of radiation and its phase ϕ are represented by non-commuting operators and that our analysis was invalidated by the uncertainty relation

$$\delta n . \delta \phi \approx 1. \tag{1.3}$$

We tried as best we could to answer all these objections and to quieten people down. We were certainly interested in seeking the truth but in raising money to build an interferometer it was desirable that our proposals should be widely regarded as sound. These difficulties about photons troubled physicists who had been brought up on particles and had not fully appreciated that the concept of a photon is not a complete picture of light. Thus many people are reluctant to accept the notion that a particular photon cannot be regarded as having identity from emission to absorption. These objections can, in fact, be answered straight out of text-books and we developed some considerable skill in expounding the orthodox paradoxical nature of light, or, if you like, explaining the incomprehensible—an activity closely, and interestingly, analogous to preaching the Athanasian Creed. In answer to the more sophisticated objection that our proposal was inconsistent with the uncertainty relation in equation (1.3) we pointed out that we were proposing to measure only the *relative* phase $(\phi_1 - \phi_2)$ between two beams of radiation; the total energy of two beams $(n_1 + n_2)$ and their relative phase $(\phi_1 - \phi_2)$ *can* be represented by commuting operators and *can* be represented classically.

Finally, there were the objections based on laboratory experiments which claimed to show that photons are not correlated at two separate detectors. Here we were on sure ground because we had already done our own careful laboratory test of the principle. In 1955, I had borrowed the dark-room which housed the spectro-heliograph at Jodrell Bank and set up our first optical interferometer. An artificial star was formed by focusing the brightest part of a high-pressure mercury arc on to a pinhole. The light from this pinhole was then divided into two beams, by a half-aluminized mirror, to illuminate two photomultipliers mounted so that their photocathodes could be optically superimposed or separated by a variable distance as seen from the pinhole. The whole system simulated the measurement by two detectors on the ground of a star with a surface temperature of about 8000 K. After the usual troubles with the equipment we observed the expected correlation and successfully measured it as a function of the separation of the two phototubes. The correlation was in reasonable agreement with theory at all separations of the detectors and this result was published (Hanbury Brown and Twiss, 1956 a) in January 1956.

We were therefore able to face with confidence objections based on two independent experiments claiming to show that there is no correlation between photons in coherent light beams. The first was performed in Budapest by Ádám, Jánossy and Varga (1955) and was published at the same time as our own test was being made. In the introduction to their paper they stated that, according to quantum theory, the pulses produced in two separate detectors illuminated by coherent light should be independent of one another. Their aim was "to investigate the validity of this prediction of quantum theory". They illuminated two photomultipliers with coherent light from a single source and also with incoherent light from separate sources, and they counted the coincidences of the pulses produced by individual photons in the two phototubes. In an observation lasting 10 hours they found no significant correlation between the arrival times of photons and they claimed that this showed that "in agreement with quantum theory, the photons of two coherent light beams are independent of each other or at least that the biggest part of such photons are independent of each other." Since the results of this experiment were welcomed by our critics as evidence that an intensity interferometer was fundamentally unsound, we took a closer look at these claims. It was at once obvious, from a quantitative analysis of the parameters of their experiment—light intensity, resolving time, etc.—that there was no hope whatever of observing correlation within 10 hours or of testing the predictions of quantum theory. In reply we published a brief note (Hanbury Brown and Twiss, 1956 c) drawing attention to the fact that in order to observe a significant correlation (three times r.m.s. noise) Ádám et al. would have had to observe for 10^{11} years— somewhat longer than the age of the Earth.

The second experimental objection was made in 1956 at the University of Western Ontario by Brannen and Ferguson (1956) just after the publication of our own laboratory work. They designed their optical system to resemble as far as possible the one we had used at Jodrell Bank. Two photomultipliers were illuminated by coherent light from a high-pressure mercury arc via a half-silvered mirror and the outputs of the two phototubes were taken, not to a linear multiplier but to a coincidence counter. They concluded that "there is no correlation (less than 0·01 per cent) between photons in coherent light rays". They added that "if such a correlation did exist it would call for a major revision of some fundamental concepts in quantum mechanics". Again, we analysed these conclusions and found that the parameters, as before, were hopelessly inadequate to allow the detection of correlation between photons within a reasonable time. In this case the essential point is that in order to achieve a practical signal-to-noise ratio with a coincidence counter one needs an intense source of light with an extremely narrow bandwidth, and this they did not have. We published a short note (Hanbury Brown and Twiss, 1956 c) showing that it would have taken Brannen and Ferguson 1000 years to observe a significant correlation.

Both these experiments were beyond reproach from an experimental point of view, but since they had been planned without an adequate theoretical foundation they were far too insensitive to be of any significant use. Nevertheless, they did provide our opponents with ammunition.

1.5 *A Pilot Model of an Optical Stellar Interferometer*

Bloody but unbowed we were ready to build a pilot model of a stellar intensity interferometer which would measure the brightest star in the sky—Sirius. This was intended to demonstrate that the method actually worked, to verify that the measurements could be made in the presence of atmospheric scintillation and to provide us with practical experience of working on a star. To save time and money we used two anti-aircraft searchlights of the largest type which I borrowed from the Army; we removed their arc lamps, substituted photomultipliers and mounted them on railway sleepers in a field at Jodrell Bank. The correlator had to be built from scratch. Details of this experiment are described in chapter 7. It proved peculiarly difficult to do because Sirius only reaches a maximum elevation of about 20° at Jodrell Bank, and it took the whole winter (1955–1956) to accumulate 18 hours of satisfactory observations. To achieve even this short exposure it was necessary to have the equipment standing by in full working order on 60 nights. Nevertheless, the experiment was a success; the apparatus was crude, but worked well enough. It gave a reasonable value for the angular diameter of Sirius (Hanbury Brown and Twiss, 1956 b, 1958 b); the measurements were certainly made through a scintillating atmosphere, and we learned some valuable practical lessons. For example, we found that it was essential to

9

screen the leads from the phototubes to the correlator with great care in order to avoid the pick-up of radio signals; we discovered that it was essential to heat exposed mirrors to avoid the formation of dew and we realized the importance of making measurements in mono-chromatic light in order to simplify the analysis of the data.

1.6 Two more Laboratory Experiments

We were now certain that a full-scale stellar intensity interferometer could be built and that it would be a practical and valuable instrument. However, to make doubly sure, we thought it worth while to repeat our original laboratory experiment with the better phototubes which had now become available and, also, in an entirely new experiment to demonstrate that correlation between individual photons could be observed with coincidence counters. The first of these experiments is described in chapter 6 and gave us what we wanted—a more precise verification of the theory. The second experiment (see chapter 6) was, of course, done in response to the many objections raised by Ádám *et al.* and Brannen and Ferguson. In both our previous laboratory experiments we had used such a large flux of photons that the output pulses overlapped in the phototubes to produce random noise and the arrival of individual photons could not be distinguished. In order to get an adequate signal-to-noise ratio in a coincidence-counting experiment in which individual photons can be counted, one needs an intense and narrow-band source of light and we therefore built an entirely new equipment using a mercury isotope lamp. This type of lamp was temporarily difficult to obtain in England, and so the experiment was performed in Sydney by Twiss and Little in 1957 (Twiss, Little and Hanbury Brown, 1957). They used an electrode-less radio-frequency discharge in mercury-198 vapour as a source of light, 1P21 phototubes and a coincidence counter with the resolving time of 3.5×10^{-9} s. They compared the coincidences between photons arriving at these two phototubes with the coherent and in-coherent illumination. In a test lasting 8 hours they found that, with coherent illumination, the number of coincidences was increased by 1.93 ± 0.17 (p.e.) per cent, which was in satisfactory agreement with the theoretical estimate of 2.07 per cent. Thus their experiment confirmed the correlation between photons in a striking manner. (Details are given in chapter 6.) A similar experiment with a similar result was carried out later by Rebka and Pound (1957) at Harvard.

1.7 Financing and Building the Narrabri Stellar Interferometer

The way was now clear to plan a practical stellar intensity interfero-meter which would measure a significant number of stars. We had worked out the theory and defended it successfully against all comers; we had also demonstrated that it was correct by three separate laboratory experiments and one measurement of a star (Hanbury Brown and

Twiss, 1956 a, b, 1957 a, b, 1958 a, b). To measure a reasonable number of stars, we knew that we should need to build an instrument with two reflectors roughly 7 m in diameter and with a maximum baseline of about 200 m. For a novel instrument this seemed rather an ambitious programme so I sought the advice of Professor P. M. S. Blackett, later Lord Blackett. He already knew about the work because, during its early stages, he was Langworthy Professor of Physics at Manchester University but by this time he had moved to Imperial College and was engaged in energizing the financing of research by the Department of Scientific and Industrial Research (D.S.I.R.). He had a great gift for encouragement once he was persuaded that a proposal was worth while, but he was not easy to convince. Nevertheless, I did convince him and he lent his support to the project, suggesting that I submit a proposal to the D.S.I.R. We now engaged in the money-raising game.

In these first discussions in 1956, I had in mind an instrument costing considerably less than £50 000 but in 1957, when we came to look more closely at the cost, we realized that this would not be enough. It seemed a lot to ask for, unless we had a partner to share the cost. Richard Twiss found the answer. He was working at the time in Sydney and discussed our proposal with Professor Harry Messel who was busy revitalizing the research of the moribund School of Physics at the University of Sydney. Harry Messel came forward with a handsome offer to share the cost—even, if necessary, to bear the whole cost. At the same time I approached my friend Herman Lindars of Dunford and Elliott in Sheffield; he had been responsible for designing and making the control desk and computer of the 250 ft radio telescope at Jodrell Bank and was interested in novel instruments. I asked him to help me to study the feasibility of the proposed instrument and to make a rough estimate of its cost. Herman Lindars generously gave his own time and that of his firm.

We kicked the ball off in February 1958 with the first proposal to the D.S.I.R. for an optical stellar intensity interferometer capable of measuring stars brighter than magnitude $+2\cdot5$; we pointed out that there were about 65 stars over the whole sky that such an instrument could measure. It was to be a joint project between the Universities of Manchester and Sydney, with the instrument being built largely in the U.K. but installed and operated in Australia. The University of Sydney undertook to meet all the costs in Australia including shipping, installation and the running costs for at least five years; furthermore, in the event of no funds being available from the U.K. they offered to contribute up to £40 000 towards the capital cost of the instrument itself. At that time we estimated that the total cost would be £70 000 and we asked the D.S.I.R. to contribute one-half.

On the whole, our proposal was well received, but one of the referees pointed out that, "what Mr Hanbury Brown is proposing is two

11

200-inch telescopes movable on railway tracks and accurately steerable so as to follow the motions of a star for hours at a time''. He maintained that it was absurd to suppose that we could produce such an elaborate set-up for the sum suggested, particularly when one remembered that the 200 in Hale telescope cost several million dollars. In the light of this opinion, the D.S.I.R., very reasonably, made a grant of £5000 for a design study leading to more precise estimates. This study was carried out by Dunford and Elliott and by the research laboratories of Mullard. Of course, the estimated cost rose, until it reached £140 000. A second proposal to D.S.I.R. in May 1959, asking them to contribute half, produced a grant of £75 000 in January 1960.

We now set to work to get the instrument built. This proved to be unnecessarily frustrating. We were working in the shadow cast by the much publicized financial difficulties of the large radio telescope at Jodrell Bank, and we had to bear the full brunt of the Civil Service's attempts to prevent another serious over-expenditure. Her Majesty's Treasury is always willing to spend a great deal of time, money and talk on saving money—provided only that the sum involved is not very large. Judging from many of the schemes, particularly for new types of aircraft, which have been financed since the last war, there is evidently a threshold above which their resistance to expenditure decreases; presumably this threshold, rather like energy levels in an atom, corresponds to a change in the administrative level at which the decision is taken. Anyway, our request for £75 000 was well below such a threshold and we had to negotiate the maximum possible number of administrative hurdles. A 'representative' supervisory committee was set up to administer the grant; every item had to be put out to competitive tender and discussed by the committee, which then had to justify their choice of tender to the D.S.I.R.; approval of the acceptance of tenders had to be given by the D.S.I.R. and so on and so on. Of course, such a rigmarole is laudable from the point of view of guarding public money, but in practice it wasted a lot of time of busy people without really sharing the responsibility. The basic responsibility in building a novel instrument is simply that the instrument, when built, shall fulfil its declared purpose; this cannot be shared by any supervisory committee and I formed the opinion that the function of financial control could have been executed more efficiently by a single responsible officer.

To cut a long and tedious story short, we did succeed in getting the component parts of the instrument built, and paid for; the steelwork was made by J. W. Ellis in Newcastle, the gears by Alfred Wiseman in Birmingham, the light-alloy framework of the reflectors by Saunders-Roe in Beaumaris and the computer and control system by Lindars Automation in Sheffield. Mullards undertook the correlator, and two other difficult components—the electro-hydraulic motors and the glass mirrors—were made by Officine Galileo in Florence. It is interesting to note that the electro-hydraulic motors, developed originally for driving guns, turned out to be extremely robust; on the other hand, the glass mirrors, having no precedent, gave us a lot of trouble.

Their design emerged only after much trial and error. At first I thought we might use glass mirrors made in the same way as some searchlights, by pressing a heated glass blank over a former. But samples made by this process were not good enough. We also had samples made by spinning aluminium but, again, they were not nearly good enough. We rejected plastic mirrors on the grounds that there was insufficient evidence about their mechanical stability and we finally decided on glass figured by conventional methods. We planned to use a mosaic of hexagonal mirrors with spherical curvature each roughly 40 cm across and we sought samples and quotations from several firms in U.K., Germany, France and Italy. I tested these samples in the underground concrete tunnel which connects the control building at Jodrell Bank with the 250 ft radio telescope. They were all of adequate optical performance but the price varied by a factor of ten. Eventually we chose those from Officine Galileo in Florence and placed an order for 540 glass mirrors all with the same nominal curvature and each at a cost of about £20. At that time it was a disappointment to me that, in order to avoid the expense of figuring two surfaces, these mirrors had to be front-aluminized and coated with silicon dioxide; I thought that they would not stand up to years of exposure out of doors. In the event, however, my fears were unfounded; after ten years of constant use the mirrors at Narrabri look almost as good as new.

All the component parts of the instrument except the correlator were completed by August 1961. Time and money prevented a proper dress rehearsal of fitting them all together before export but the two reflector frameworks were assembled among a lot of boats on the slipway of Saunders-Roe at Beaumaris and given a few hasty tests. After some urgent last-minute modifications, conducted almost entirely in pouring rain, they were put into crates and shipped to sunny Australia. We had to simulate the weight of the mirrors in these tests by attaching pieces of lead to the framework because we were unable to import the mirrors themselves from Italy without paying Customs Duty, and no-one had the time to argue the case with the appropriate department. The mirrors were sent direct to Australia from Italy where, as it turned out, we had a worse fight with the Australian Customs. They demanded payment on the grounds that the British Custom authorities had advised them that the mirrors could have been made in U.K.; Italian mirrors were therefore dutiable under Commonwealth agreements. After a good deal of agitation on our part they let the mirrors through without payment so that we could get on with the job of their installation, but it took us a year to convince them that we should not pay them any money.

1.8 *Installation of the Interferometer at Narrabri*

The whole interferometer, minus the correlator, arrived in Narrabri in January 1962 and I had my first view of the site which had been chosen by Richard Twiss and Harry Messel. Narrabri is a small country town on the River Namoi in northern New South Wales about 340 miles from Sydney by road. The Observatory is 12 miles from town, 15 from the airport, and is sited by kind permission of the owners on a property of about 3000 acres. Like most of the area, the site itself is about 600 ft above sea-level, flat, with few trees but with a magnificent view of the Nandewar hills which rise to 5000 ft about 20 miles to the east. The climate is beautiful in winter but too hot in summer. The average annual rainfall is 24 in. but the rain comes irregularly and in very large doses. About 60 per cent of all nights are clear and the moonless sky is dark; for more than half the time the extinction is low but on several occasions during the last ten years our work has been seriously hampered by dust.

When I arrived early in 1962 the construction of the site was already completed under the supervision of Malcolm Nicklin of MacDonald, Wagner & Priddle; this had been negotiated by Cyril Hazard, one of my colleagues at Jodrell Bank who had earlier transferred to Sydney University in 1961. Towards the end of 1961 he was joined by John Davis, also from Jodrell Bank. The site consisted of a circular track, 188 m in diameter, a central mast to carry the cables, an enormous and expensive garage for the reflectors, a control building (air-conditioned) and three small prefabricated houses for the observers. The indigenous wire fencing of the outback protected it from sheep, cattle, kangaroos and emus, and it was connected to civilization by several miles of telephone and electricity supply lines and an ungravelled dirt road. All that remained to do was for the three of us to put the bits of the actual interferometer together with the help of a small team of engineers from Sydney, two engineers from Sheffield and the local know-how of Graham Gifford, our indispensable caretaker-mechanic.

This proved to be an exceptionally trying experience which I would not care to repeat. The whole operation had to be done as fast as possible because we could not afford to keep the engineers for long. The weather did not help; heavy rain reduced the surrounding black soil to a quagmire which could only be negotiated in a Land-Rover and between the storms it was extremely hot with the famous Australian Bush Fly out in force. To make matters worse, the instrument had never been put together before in its entirety and it is not easy to make vital modifications when the nearest workshop is over 100 miles away. One rather comic example of our troubles was that the two reflectors would not fit into their splendid and costly garage; the dimensions of the building had been specified for reflectors standing straight on their tracks but the inevitable curvature of the rails rotated them slightly and their corners stuck out of the doors. Fortunately we were able to cure this miscalculation by simply chopping off the parts that stuck out.

Other troubles were not so easily solved. One of the worst of these concerned the mirrors which had arrived from Florence, beautifully packed and coated with a protective layer of plastic. A small party, which consisted of every able-bodied person within sight, spent two months assembling the mounts and putting the mirrors on the framework of the reflectors. When it was all done we celebrated the occasion by stripping the protective coat off the mirrors. In 140 cases the plastic coating brought some of the mirror surface with it, damaging it so badly that the mirrors were ruined. Very sadly we spent another month dismantling the 100 worst cases and returned them to Florence for re-coating. The trouble was probably due to the extreme heat, as the temperatures were well over 100°F for a good deal of the time.

As fast as we cured one trouble another appeared. The reflectors, finally complete with all their mirrors, were ready for test in October 1962. We pointed them horizontally at a distant gum tree on which we had mounted a lamp and examined the 'image' on a sheet of ground glass. Each of the 252 mirrors on each reflector was then adjusted individually to give a circular patch of light about 13 mm in diameter. We then tracked Jupiter over a wide range of elevations and photographed the 'image' in a remotely controlled camera at the focus. To our disappointment the photographs showed that in both reflectors the size and shape of the image varied greatly with elevation; the circular patch deformed into an ellipse with its major axis in the azimuthal plane. At 70° elevation the major axis of this ellipse increased to no less than 60 mm. After a good deal of anxious research the cause of this trouble was traced to a bending of the main steel tube which carries the framework on which the mirrors are mounted. As nothing simple could be done to strengthen this tube, we accepted the larger image and developed a systematic method of aligning the mirrors to give the minimum size of image over the working range of elevations. The final alignment was evolved by measuring the deflection of each mirror with the aid of a television camera mounted at the focus. This television system was also used to monitor the image while tracking Jupiter and to make the final adjustments of the time constant and servo-gain of the photoelectric star-guiders.

Another significant technical problem which showed up at this point was the difficulty of getting the two reflectors to roll smoothly and precisely on the circular railway track. This proved surprisingly hard to do. We had to replace the original wheels of mild steel with new ones of very hard steel and then machine them into sections of a cone with its apex at the centre of the 188 m circle. The axles had also to be aligned optically on the centre of the track with extreme precision, and the track itself had to be adjusted to make it level.

These, and many other mechanical problems, kept us and the Sheffield team fully occupied until January 1963 when the electronic correlator arrived, bringing with it its own train of peculiarly difficult electronic problems. Its installation took four months using the combined ingenuity of Arthur Browne of Mullards who accompanied it and of our own team which was now joined by Roy Allen of Jodrell in March 1963. Not until May 1963 was the complete interferometer ready to be tested for the first time on the bright star β Centauri.

1.9 *Preliminary Test, Pilot Programme and Teething Troubles*

This was the long-awaited moment; over ten years work had gone into its preparation and, short of an earthquake, one felt there was nothing left that could possibly go wrong. Everything worked well, even the star-guiding, but there was absolutely no correlation. I must admit to a slight sense of panic; a hasty review of the whole theory with Richard Twiss was followed by a critical examination of every nut and bolt. We found an explanation fairly quickly; we had equalized all the time delays in the cables and the correlator, but not in the phototubes. Urgent experiments with short pulses of light from a spark showed that the delays in the two phototubes were different and must be equalized. This done, our observations of β Centauri were more successful but there was still only half the expected correlation. A second anxious scrutiny failed to explain the trouble until, as a last resort, we turned the equipment on to Vega. All became clear; by sheer bad luck we had chosen to start with a star of unknown complications. At that time, β Centauri was thought to be a widely spaced binary with a relatively faint secondary component. We later showed that there are three stars, the 'primary' itself consisting of two stars of comparable brightness which had not previously been resolved, and it was this third star which explained the low correlation. Greatly relieved, we persevered with the measurements on Vega and completed them successfully in August 1963. We published the results as a preliminary paper, to show that the instrument actually worked (Hanbury Brown, Hazard, Davis and Allen, 1964). After a few minor improvements to the equipment we then embarked on the measurement of four more stars in a programme lasting about 250 hours during 1964.

This pilot programme showed that the limiting sensitivity of the equipment was about one magnitude brighter than we had hoped. (We later realized that a significant part of this loss was due to excess atmospheric dust following the eruption of the volcano Mount Agung in Bali in 1963.) It also showed that the stability and reliability of the electronic correlator were not good enough. In fact the design of a satisfactory correlator proved to be by far the most difficult problem of the whole project. It is described in Chapter 8. The basic difficulty is to free the output from small irregular zero-drifts. In this respect the most critical component is the linear multiplier, and it was not until the wideband transistorized linear multiplier designed by R. H. Frater of the School of Electrical Engineering in Sydney became available that we could stabilize the correlator satisfactorily. We introduced this multiplier in December 1964 and, at the same time, Mullards Ltd. presented us with a completely transistorized programme unit which greatly improved the reliability of the correlator. Also I visited R.C.A. in Lancaster, Pennsylvania, and persuaded them to make us some better phototubes.

Following this pilot programme of 1964 we worked out a standard procedure for observing stars and analysing the results. In doing this we were fortunate to have the help of an astonishingly industrious and critical visitor, Professor Dan Popper of the University of California. We have stuck religiously to these procedures ever since, so that all our results have been taken and analysed in the same way.

As was to be expected with such a novel instrument, technical troubles were accompanied by financial and administrative problems. When I first arrived in Sydney in 1962 it was clear that, as in all new projects (e.g. the Opera House in Sydney), we had underestimated costs. We had expected that the total installed cost would be £140 000 sterling. In fact it was £225 000 sterling, the principal discrepancy being in the costs of construction of the site and more particularly in the labour costs of wiring and erecting the equipment. The D.S.I.R. had contributed £93 000 and the School of Physics at Sydney wished to limit their contribution to £96 000. This meant that I had to find the difference of £36 000 from somewhere else. I made unsuccessful approaches to several foundations in the U.S.A. and to the Nuffield Foundation in the U.K. Finally I went to Washington and called on the Office of Scientific Research of the United States Air Force. They were enthusiastic about the project and in January 1963 they made a grant of $122 500 to the capital cost and also a substantial contribution to the annual running costs. From that day to the end of the programme we have never been short of money.

Administration of a joint project in which one partner is 12 000 miles away and the other 350 miles from the scene of action easily gets into a three-cornered muddle of divided responsibility and ambiguous communication. Not surprisingly there were misunderstandings between the Universities of Manchester and Sydney as to who was responsible for what. They stemmed largely from the fact that the equipment was not properly tested before it left U.K. and that the cost of preparing the site in Australia was badly underestimated. I tried to reduce the friction by writing detailed accounts of what was happening and where the money was going, but in those early days I was more anxious to straighten out the practical problems at Narrabri than to apportion blame. Although it is one of the sacred cows of university dogma that teaching and research must go hand in hand, the organization and facilities of day-to-day university life are seldom designed to make it easy to do research on the actual campus, let alone beyond the black stump. In New South Wales this is made even

17

worse by the absurd centralization of population, outlook, and every-thing else in Sydney. I remember trying to explain to the Accounts department of the University of Sydney that our casual labour at the Observatory, aborigines living in tents, could not read the elaborate tax forms which they kept on sending to me and did not appear to fit into any of the social categories listed on them. Manchester and Sydney are no exception to the rule that universities tend to use too many committees and too few technicians. In practice this meant that if we wanted anything done—such as repairing the correlator or the air-conditioning or getting frogs out of the plumbing—we did it our-selves; this left me too busy to conduct an effective public relations programme. However, in the course of time all was sweetness and light due in large part, I feel, to the patience of Professor Messel and the forbearance of the Officers of Manchester University.

1.10   *The Main Work of the Observatory*

The main programme of the Observatory began with β Crucis in May 1965 and ended with δ Canis Majoris in February 1972. Broadly speaking the work had three main objectives; first of all we wanted to make a significant contribution to stellar astronomy by measuring the angular sizes of 32 single stars carefully chosen to represent the spectral range O to F. Secondly, we aimed to explore various other applications of an intensity interferometer to astronomy including the detection of close-spaced binary stars, the measurement of the angular size of an emission region surrounding a hot star, the investiga-tion of the effects of limb-darkening, polarization and rotation on a single star, and, what turned out to be the most interesting application of all, the determination of all the parameters of a spectroscopic binary. Thirdly, we were interested in developing the technique itself; in the course of doing this we examined the effects of atmospheric scintillation and Čerenkov light from the sky; we developed a reliable and stable correlator and we improved the sensitivity of the equipment. It is satisfactory to note that by the end of 1966 we had reached a limiting magnitude of $+2\cdot5$ for which the instrument was originally designed. An account of all this work is given in later chapters.

It was a lengthy and ambitious programme for such a small and novel observatory in a rather isolated place, and I am reminded of a remark I heard Sir Edward Appleton make in 1952 during a visit of U.R.S.I. to Australia. He emphasized that the most outstanding thing he remembered about his long career in research was the actual physical effort involved. I was surprised at the time, but now, 20 years later, I (and I suspect my colleagues also) understand exactly what he meant. The personal difficulties of the staff in leading a double life by trying to combine their teaching duties with a demanding research programme 350 miles away are obvious; we owe much to John Davis who for years undertook the tricky and unrewarding chore of arranging

18

the observing roster.   Moreover I am keenly aware that his job was made harder by my wish to keep the number of staff to a minimum; at no time did we have more than four senior people who could take charge of the observing and the overall number of people in the whole department at any one time never exceeded ten.   From previous experience of the early days of radar at Bawdsey Manor and of the early days of radio-astronomy at Jodrell Bank, I know how much easier it is to maintain interest in a project if everybody has a large personal share in the responsibility of the venture as well as in the routine.   This necessarily means a small group—perhaps our group was too small, but I was acutely aware that everybody's share might be in failure rather than success and that we all risked, at the least, wasting our time. However, I do believe that all of us at Narrabri felt a personal concern for the project and shared in the satisfaction of its ultimate success.

During the seven years of the main observing programme we spent at least 2500 hours actually taking measurements on stars, which means that the whole equipment ran for very much longer.   But much of our time was devoted—and that is an appropriate word—to main-taining this complicated apparatus in constant working order out in the Australian Bush.   Although Graham Gifford, our caretaker mechanic, has taken good care of the site and its machinery for over ten years, we have not had a full-time electronic technician at the Observatory. It is in fact difficult to provide a satisfactory job at such a place for someone who is sufficiently skilled to maintain and repair the correlator, computer and control system.   Each senior observer therefore had to be able to maintain the electronics himself.   This severely limited our choice of observers and made the equipment unusable by visitors working on their own.

In some ways the bush was helpful.   The dryness and clean air reduced corrosion and accounted for the surprising fact that, even after ten years, the mirrors were almost as good as new.   On the whole, however, the isolation works against efficient maintenance.   For example, spare parts take a long time to come and, on a modest budget, one cannot manage to keep everything needed in stock even if one could foresee everything necessary.   We had not expected to suffer from destructive birds.   Yellow-throated Miners (not to be confused with the Indian Mynahs), for example, are so fascinated by their own reflection that they will peck away at a mirror until its surface is ruined. This only happens when the mirrors are in the garage, and for a long time it was cured by a hawk that nested in the roof.   One is reminded of the advice given by the Duke of Wellington to Queen Victoria when she asked him how to keep sparrows out of the Crystal Palace.   Another more serious trouble was that the beautiful pink and grey parrots, galahs, loved to hang upside down on the catenary wires and peck at the smaller cables like wire-cutters until they actually severed them. We eventually discovered that the cables could be protected by wrapping them in tarred roofing felt for which, apparently, parrots have no appetite.

Other unforeseen snags arose from the changing pattern of country life. In 1962 the land was devoted to grazing, but times have changed. Sheep are no longer profitable, wheat and cotton have replaced much of the grassland and the emus and kangaroos have gone. Undesirable features followed; dust from the increasing number of ploughed paddocks, brilliant headlights from night ploughing and crop-dusting, and worst of all, the smoke made by burning-off the wheat stubble after the harvest. Soon after I arrived, television also reached the north-west plains of N.S.W. and I had to negotiate tactfully in Melbourne a redistribution of transmitting frequencies so as to avoid serious radio interference with the operation of our correlator.

A pleasant surprise was the cordial and tolerant attitude of the local inhabitants among whom I would like to mention Mr. and Mrs. L. P. Miller who run the property on which the Observatory is built. We appreciated, too, the interest of the hundreds of people who ventured to come and see us. Many of course, were astronomers from all over the world, but we also entertained a wide variety of other visitors. On the one hand we have been honoured by visits from the Governor General of Australia, and from London, the President of the Royal Society; on the other hand, there have been countless school-children and the inevitable bus loads of tourists, most of whom, and I sympathize, have got out of the bus largely to stretch their legs.

Looking back on the whole programme I am satisfied that the work of these years did achieve our three principal objectives. We made reasonably precise and reliable measurements of 32 single stars; these include the first measurements ever made of a main sequence star and the first measurements of any star earlier than type M. We increased the number of known angular diameters from 6 to 38 and we showed how this number could be increased indefinitely. This work stands as a permanent and valuable contribution to stellar astronomy and it alone justifies the time, anxiety and expense of the whole project.

Our demonstrations of the potentialities of an intensity interferometer were necessarily less satisfactory. They were, of course, limited by the fact that we were trying to show with a relatively small instrument what a much larger one could do. The experiments on limb darkening, stellar rotation and polarization were all too close to the limits set by signal-to-noise ratio; furthermore, much to my disappointment, the sensitivity of the Narrabri interferometer is too low to observe the pulsations in radius of a Cepheid variable, in my view the most interesting untried application of an interferometer. However, these shortcomings are largely redeemed by our successful observations of the spectroscopic binary α Virginis; the measurement of the orbital inclination and the *distance* of this star are most striking examples of the potential of an interferometer.

Lastly I think we made a reasonable, though not comprehensive, job of developing the technique itself; but we did solve the most difficult problem of all—that of building a stable and reliable correlator. As a result it is now possible to take the next step in this work with confidence and to tackle the exciting possibilities which are outlined in the last chapter of this book.

1.11 *The Future*

Most active research groups, not only in astronomy, eventually reach the conclusion that what they really need is a larger and more expensive instrument. We at Narrabri Observatory are no exception and we have proposed (Chapter 12) that the next step should be to build a larger and more sensitive interferometer capable of measuring stars $4\frac{1}{2}$ magnitudes (~ 60 times) fainter than we can now reach. Broadly speaking, we propose that having solved the difficult and classical problem of how to measure the angular size of the stars, we should now explore thoroughly the possibilities of this new technique. In observational astronomy, as in most branches of science, it is the introduction of new tools such as the photometer, the spectroscope, the radio-telescope and the image intensifier, which has made significant advances and guided the direction of research. The intensity interferometer is a completely new tool and as such, it can be expected to contribute qualitatively new information to advance our understanding of the stars.

In Chapter 12 I have outlined the more obvious research programmes which could be undertaken. The first of these (§ 12.2) is the measurement of the emergent fluxes, effective temperatures and radii of single stars and is a straightforward extension of the work at Narrabri to a greater variety of stars. We are quite confident that this would yield worth while results and it is, in effect, the bread and butter of our proposal. However, some of the other programmes such as the observation of the radial pulsations of Cepheid variables (§ 12.4) or the shape of rotating stars (§ 12.5) are speculative and might well turn up surprises—we cannot tell until we have tried.

Our current efforts to raise money for a new interferometer began in earnest in 1971 when we submitted a proposal to the Australian Government. We asked for a grant towards a design study of a large stellar intensity interferometer (see Chapter 12) estimated to cost about as much as a 2 m telescope with dome and accessories—in those days about $A 2 m. Since then, our proposal has been assessed and discussed around the world by astronomers, weighed in the scales against scientific fashion and relevance by laymen and politicians, and has come through with flying colours. It has even survived two elections of the Australian Government. In my own view, the Slough of Despond, the Hill Difficulty and Doubting Castle lie behind us. I am hopeful—and if you remember your Bunyan—Hopeful was not disappointed.

21

CHAPTER 2

a simple explanation of how an intensity interferometer works

2.1 *Introduction*

Experience suggests that many people who open this book will be looking for an explanation of intensity interferometry and hoping for a satisfactory mental picture of how an intensity interferometer 'actually works'. There are two difficulties in presenting such a picture; first, there is no satisfactory mental picture of the wave–particle theory of light; secondly, most physicists and astronomers are not familiar with what takes place when a wave with the characteristics of random noise is demodulated by a square-law detector—they feel uneasy or baffled when confronted with an explanation which is acceptable to a radio engineer. An obvious remedy is to describe it in terms of photons; however, as remarked later, explanations in terms of photons are liable to serious misinterpretation. For this reason the explanation offered in this chapter does involve some understanding of what happens in a square-law detector.

To put the intensity interferometer into a better perspective we shall first discuss Michelson's stellar interferometer. Since there are many descriptions of this classical instrument in standard texts, the discussion will be brief.

2.2 *Michelson's Stellar Interferometer*

Figure 2.1 illustrates the principle of Michelson's interferometer. Light from a star is received on two small separated mirrors M_1, M_2 and is reflected via M_3, M_4 into the larger mirror M which brings the two beams together in the focal plane at O. By this arrangement the two images of the star, as seen in the two small mirrors, are superimposed. If the two mirrors are not too far apart the two images interfere, forming alternate bright and dark bands across the patch of light (fig. 2.2). When the two mirrors are close together, or the star has a small angular size, then the intensity in the dark bands approaches zero and the contrast or *visibility* of the fringes is high; as the mirrors are separated this visibility decreases until eventually the fringes vanish.

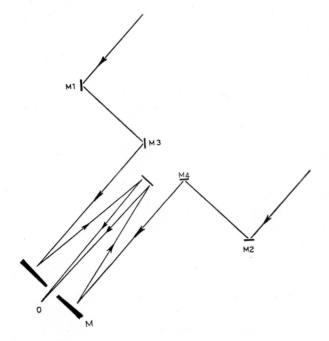

Fig. 2.1.   Michelson's stellar interferometer.

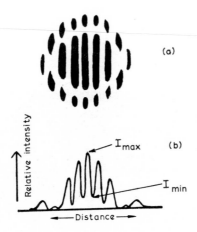

Fig. 2.2.   Fringes formed in a Michelson interferometer.

23

If the visibility at any mirror spacing d is defined by

$$V_d = (I_{max} - I_{min})/(I_{max} + I_{min})$$

where I_{max} and I_{min} are the maximum and minimum intensities in the fringes, then it can be shown (e.g. Born and Wolf, 1959) that

$$V_d = |\gamma_d| \tag{2.1}$$

where $|\gamma_d|$ is the *degree of coherence* (§ 3.2) of the light at the two mirrors, which depends upon the angular diameter θ of the star, the wavelength of light and the separation between the mirrors d. For a star with a uniform circular disc, V_d varies with d as shown in fig. 2.3 and reaches zero when

$$d = 1 \cdot 22\lambda/\theta. \tag{2.2}$$

Michelson's stellar interferometer was mounted on the 100 in telescope at Mount Wilson. The separation between the two small mirrors M_1, M_2 could be controlled by the observer and the maximum possible value was 6 m. In operation the fringes were first observed with the two mirrors close together and then the spacing was increased until they disappeared. From a measurement of this critical spacing the angular diameter of the star was found from equation (2.2).

Michelson's interferometer has serious disadvantages. First, in order that there should be steady fringes, it is essential that any difference between the path lengths from M_1, M_2 to O should be extremely stable. Any change in the difference, comparable with the wavelength

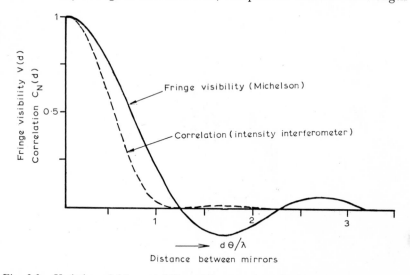

Fig. 2.3. Variation of fringe visibility with separation of mirrors for a Michelson interferometer (full line), an intensity interferometer (broken line). The source is a uniform circular disc.

24

of light, will displace the fringes in the focal plane. Slow displacements are relatively unimportant but rapid displacements blur the fringes and reduce their apparent visibility. A second requirement is that the actual difference between the two paths must be small compared with the coherence length (§ 3.4) of the light. Both these requirements demand an extremely stable and rigid structure which must be pointed precisely at the star and, for these reasons alone, it is difficult to construct and operate a large instrument. Finally, the most serious trouble is that atmospheric scintillations introduce rapid, random and uncorrelated differences into the paths of the light reaching the two small mirrors. In consequence the measurements of fringe visibility are significantly affected by the atmospheric 'seeing' and have proved to be variable (Pease, 1931).

These difficulties—the need for extreme mechanical precision and the dependence of the measurements on the 'seeing'—have so far prevented the development of Michelson's stellar interferometer beyond the original 6 m model. An attempt was made by Pease in the period 1925–1937 (Pease, 1925, 1930) to extend the work by building a 15 m interferometer at Mount Wilson. However, this larger instrument proved extremely difficult to operate and would not yield consistent results. Thus the list of stars measured by Michelson's interferometer still comprises only the six stars shown in Table 2.1; these are all cool giants or super-giants in the spectral range K to M and do not include a single member of the main sequence.

Star	Spectral type	Luminosity class	Angular diameter $\times 10^{-3}$ seconds of arc
α Boo	K2	Giant	20
α Tau	K5	Giant	20
α Sco	M1–M2	Super-giant	40
β Peg	M2	Giant	21
σ Cet	M6e	Giant	47
α Ori	M1–M2	Super-giant variable	34→47

Table 2.1. Stars measured with Michelson's interferometer. From Pease (1931).

2.3 The Intensity Interferometer

A simplified outline of an intensity interferometer is shown in fig. 2.4. Light from a star is received on two separated photoelectric detectors, D_1, D_2 and produces output currents i_1, i_2. These currents fluctuate and an intensity interferometer relies on the fact that the fluctuations are partially correlated. The principal component of the fluctuations is the classical *shot noise* associated with any current but,

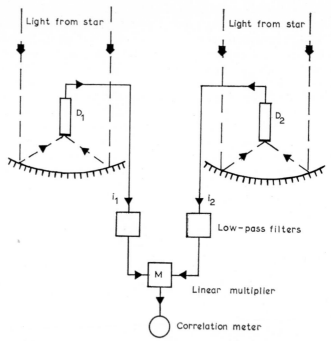

Fig. 2.4. Simplified outline of an intensity interferometer.

in addition, there is a smaller component called *wave noise* which corresponds to the fluctuations in the intensity of the light wave. One can think of this wave noise as the envelope of the light wave rectified by the photoelectric detector. The major component of noise, that is to say the shot noise, is not correlated with the shot noise in the other detector; but the minor component, the wave noise, is correlated with the wave noise in the other detector provided that there is some degree of coherence between the light at the two detectors. Thus, when the two fluctuating currents are multiplied together in the linear multiplier M, there will be a small positive product corresponding to the correlation of the wave noise components. It can be shown (§ 3.5) that this product, or correlation $c(d)$, is proportional to the square of the degree of coherence of the light at the two detectors and is therefore also proportional to the square of the fringe visibility which would be observed in a Michelson interferometer under the same conditions. Thus, if $\Delta i_1(t)$, $\Delta i_2(t)$ are the fluctuations in the two currents, we may write

$$c_N(d)/c_N(o) = \langle \Delta i_1(t)\Delta i_2(t)\rangle/(\langle i_1\rangle\langle i_2\rangle) = |\gamma_d|^2 = V_d^2 \qquad (2.3)$$

where $|\gamma_d^2|$ is the degree of coherence, V_d is the corresponding fringe visibility, $c_N(d)$ is the normalized (§ 10.1) correlation with a baseline d and $c_N(o)$ is, in effect, a constant of the equipment.

26

It follows that, when observing a star, the correlation will decrease with increasing baseline as shown by the broken line in fig. 2.3 and that a measurement of this curve will give the angular size of the star. It should be remarked that there are no interference fringes formed in an intensity interferometer and in a classical sense there is no interference of light; the interference takes place between the electrical fluctuations at the linear multiplier. It may also be noted that, unlike a Michelson interferometer, the instrument measures the *square of the modulus* of the complex degree of coherence and so the phase of this complex function is lost. Broadly speaking, this means that one cannot reconstruct the angular distribution across an asymmetrical source without ambiguity; for example, when observing a double star with two unequal components, one cannot tell which star is 'on the left' and which is 'on the right'.

We shall now enquire why the wave noise in the two detectors is correlated, since this is the principal question which many people have asked and found difficult to understand. One way of looking at the problem is as follows. Consider two elementary points P_1, P_2 in fig. 2.5

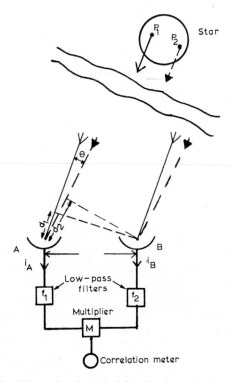

Fig. 2.5. Illustrating the principle of an intensity interferometer.

27

on the surface of a star. Each point radiates white light and is completely independent of any other point. Following conventional Fourier analysis we may represent the wave form of this light as the superposition of a large number of sinusoidal components, each component having a steady amplitude and phase over the period of observation but both the amplitude and phase being random with respect to the other components. We shall assume that the light is limited in bandwidth by an interference filter in front of each detector to some convenient band, say 450 nm $\pm$ 5 nm and that the electrical filters f_1, f_2 pass all frequencies in the band 1–100 MHz. Consider now one Fourier component of the light $E_1 \sin(\omega_1 t + \phi_1)$ reaching detector A from the point P_1, and a second component of *different* frequency $E_2 \sin(\omega_2 t + \phi_2)$ in the light reaching A from P_2. The output current from A is proportional to the *intensity* of the light and so we may write, assuming simple linear polarization,

$$i_A = K_A[E_1 \sin(\omega_1 t + \phi_1) + E_2 \sin(\omega_2 t + \phi_2)]^2 \qquad (2.4)$$

where K_A is a constant of the detector.

The same two Fourier components will also illuminate B and give rise to a current

$$i_B = K_B[E_1 \sin(\omega_1(t + d_1/c) + \phi_1) + E_2 \sin(\omega_2(t + d_2/c) + \phi_2)]^2. \qquad (2.5)$$

Expansion of equations (2.4) and (2.5) shows that in the output of each detector there are four components:

$$i_A = \tfrac{1}{2}K_A\{(E_1^2 + E_2^2) - [E_1^2 \cos 2(\omega_1 t + \phi_1) + E_2^2 \cos 2(\omega_2 t + \phi_2)]$$
$$- 2E_1 E_2 \cos[(\omega_1 + \omega_2)t + (\phi_1 + \phi_2)]$$
$$+ 2E_1 E_2 \cos[(\omega_1 - \omega_2)t + (\phi_1 - \phi_2)]\} \qquad (2.6)$$

$$i_B = \tfrac{1}{2}K_B\{(E_1^2 + E_2^2) - [E_1^2 \cos 2(\omega_1(t + d_1/c) + \phi_1)$$
$$+ E_2^2 \cos 2(\omega_2(t + d_2/c) + \phi_2)]$$
$$- 2E_1 E_2 \cos((\omega_1 + \omega_2)t + \omega_1 d_1/c + \omega_2 d/c + (\phi_1 + \phi_2))$$
$$+ 2E_1 E_2 \cos((\omega_1 - \omega_2)t + \omega_1 d_1/c - \omega_2 d_2/c + (\phi_1 - \phi_2))\}. \qquad (2.7)$$

The first term in both of these equations is the familiar d.c. component proportional to the total light flux falling on the detector. In a practical instrument this is rejected by the filters (f_1, f_2 in fig. 2.5) and is measured in a separate circuit as a record of the light flux from the star. The second and third terms correspond to second harmonics ($2\omega_1, 2\omega_2$) and sum frequencies ($\omega_1 + \omega_2$) of the light respectively; quite apart from worrying about their physical significance, we may dismiss them because they do not lie within the frequency range passed by the filters. It is the fourth term, corresponding to *difference frequencies* of the form $(\omega_1 - \omega_2)$ which concerns us here. If these difference frequencies lie

28

within the pass-band of our filters, 1–100 MHz, they will reach the multiplier and there will therefore be two components at the multiplier of the form

$$i_A = K_A E_1 E_2 [\cos [(\omega_1 - \omega_2)t + (\phi_1 - \phi_2)]] \qquad (2.8)$$

$$i_B = K_B E_1 E_2 [\cos [(\omega_1 - \omega_2)t + (\phi_1 - \phi_2) + \omega_1 d_1/c - \omega_2 d_2/c]. \qquad (2.9)$$

One can see from these two equations that these two components (i_A, i_B) are correlated; they have the same frequency but they differ in phase by $(\omega_1 d_1/c - \omega_2 d_2/c)$. Their product, or correlation, is

$$c(d) = K_A K_B E_1^2 E_2^2 \cos [(\omega/c)(d_1 - d_2)] \qquad (2.10)$$

where, for simplicity, we have put $\omega_1 \approx \omega_2 = \omega$. It is important to note that the phase difference between these correlated components (i_A, i_B) is *not* simply the phase difference of the light waves at the two detectors but is the *difference* between the relative phases of the two Fourier components at the detectors. Finally, by simple geometry, we may rewrite equation (2.10) as

$$c(d) = K_A K_B E_1^2 E_2^2 \cos (2\pi d\theta/\lambda) \qquad (2.11)$$

where d is the separation or baseline between the detectors, θ is the angular separation of the two points P_1, P_2 on the star and λ is the mean wavelength of the light.

To extend this result to give the total correlation observed from a star, it is necessary to integrate equation (2.11) over all possible pairs of points on the disc of the star, over all possible pairs of Fourier components which lie within the optical bandpass and over all difference frequencies which lie within the bandpass of the electrical filters. Such an integration (Hanbury Brown and Twiss, 1957 b) yields the strikingly simple result, quoted in equation (2.3), that the correlation is proportional to the square of the modulus of the complex degree of coherence $|\gamma|^2$ of the light at the two detectors. It is therefore also proportional to the square of the fringe visibility in a Michelson interferometer with the same baseline. It follows that we can measure the angular size of a star by measuring the correlation $c(d)$ as a function of the separation d between the detectors.

The principal advantage of an intensity interferometer can now be understood in terms of this 'Fourier component' model. The essential point is that the correlation is a function of the difference in phase between the low-frequency beats formed at the two detectors. It is *not* a function of the phase differences of the light waves at these points and so we are not concerned, as in Michelson's interferometer, to ensure that any path differences are less than the wavelength of light. Thus we may, if we wish, delay the light reaching one detector by several thousand wavelengths without affecting the correlation, provided that this delay is small compared with the period of the highest beat-frequency which we pass to the multiplier. The reason for this remarkable property of an intensity interferometer is clear from the model; when we delay the light by a time τ both Fourier components ω_1 and ω_2 are delayed equally and so their beat-frequency $(\omega_1 - \omega_2)$ is simply delayed by a time τ. If the highest beat-frequency ($\sim 10^8$ Hz) passed to the multiplier is roughly one million times less than that of the light ($\sim 10^{14}$ Hz), then a path difference of some thousands of wavelengths will have a negligible effect on the relative phases of the two beat-frequencies at the multiplier, and hence will not affect the correlation. It is also interesting to note that it makes no difference whether the delay is inserted in the light path before detection or afterwards in the electrical path between the detector and the multiplier. Perhaps one should also note at this point that these statements about the effects of a delay are only true provided that the delay itself is non-dispersive over the relevant optical or electrical bandwidth.

It follows from this discussion that the use of intensity interference completely transforms the problems of building an interferometer. Consider first the problem of achieving the necessary mechanical precision which is such a serious obstacle in Michelson's interferometer. As we have seen, in an intensity interferometer we must ensure that any path differences between the two arms of the instrument are small compared with the wavelength of the highest beat-frequency which reaches the multiplier. If therefore we restrict this frequency to 100 MHz, then we need only equalize the two paths with a precision of about 30 cm, and this is simple to achieve even in a very large instrument. The second major problem, the effects of atmospheric scintillation, is overcome for the same reason. Thus it can be shown that any differential path length or time delay introduced by atmospheric irregularities is likely to be very much less than 30 cm and it follows that an intensity interferometer is almost completely insensitive to the effects of the atmosphere.

To summarize, it is possible to build an intensity interferometer with the very long baselines and extremely high resolving powers which are needed to measure a reasonable sample of stars, particularly of hot stars. It is also possible to make reliable measurements through the Earth's atmosphere.

Finally, it should be said that there are many alternative ways of understanding how an intensity interferometer works. For example, one may follow the radio-engineer and regard the incident light as band-limited Gaussian noise which is demodulated by a square-law detector. If the incident light is a plane wave, then the envelope of this wave is identical at two points separated by a baseline parallel to the wave front. It is then clear that the demodulated components of the envelope in the output currents of the two detectors must be correlated, and this correlation can be evaluated by conventional formulae from the theory of square-law detectors. As another alternative one can follow some physicists and explain the operation in terms of Bose–Einstein statistics and the 'bunching' of photons in the same cell of phase-space. This bunching is illustrated in fig. 4.8 in chapter 4 which shows the conditional probability that a second photon is received by a detector in a short time interval $\Delta\nu\tau$ after receipt of a first photon. The curve assumes that the light beam is a plane polarized wave with a Gaussian spectral profile of bandwidth $\Delta\nu$ and the conditional probability has been normalized to that for a stream of independent particles. It is interesting to note that, for very short intervals, the probability of finding a second photon is twice that for independent particles. If the two detectors are effectively in the same cell of phase-space, for example, if they are separated parallel to the wave front, then the arrival of photons will appear to be correlated in time; they will tend to arrive in pairs. The operation of an intensity interferometer can certainly be interpreted in this way, nevertheless it is my experience that the semi-classical 'Fourier component' model, as presented in this chapter, is freer from conceptual traps and is a more effective tool for getting the correct answers to detailed quantitative problems about an intensity interferometer.

CHAPTER 3

the·theory of coherent light

3.1 *A Mathematical Description of Light*

In several classical papers (e.g. Einstein and Hopf, 1910; Einstein, 1915; von Laue, 1915 a, b) it has been established that white light of thermal origin has the properties of a Gaussian random process.   We may represent, for example, the electric vector of a light wave as the superposition of a set of Fourier components of different frequency with amplitudes and phases which are statistically independent and randomly distributed.   If, therefore, $V^{(r)}(t)\,(-\infty\leqslant t\leqslant\infty)$ is a Cartesian component of the electric vector at a fixed point in space, then we may represent it by the Fourier integral

$$V^{(r)}(t)=\int_0^\infty a(\nu)\cos\left[\phi(\nu)-2\pi\nu t\right]\mathrm{d}\nu. \tag{3.1}$$

However, following most modern texts on optical interferometers, e.g. Born and Wolf (1959), Françon (1966), Steel (1967), we shall transform this representation into the complex *analytic signal* $V(t)$ by associating with $V^{(r)}(t)$ the conjugate function $V^{(i)}(t)$ so that

$$V(t)=V^{(r)}(t)+\mathrm{i}V^{(i)}(t) \tag{3.2}$$

where

$$V^{(i)}(t)=\int_0^\infty a(\nu)\sin\left[\phi(\nu)-2\pi\nu t\right]\mathrm{d}\nu \tag{3.3}$$

and we may then write the analytic signal as

$$V(t)=\int_0^\infty a(\nu)\exp\mathrm{i}\left[\phi(\nu)-2\pi\nu t\right]\mathrm{d}\nu \tag{3.4}$$

Alternatively, if $V^{(r)}(t)$ is represented by a Fourier integral of the form

$$V^{(r)}(t)=\int_{-\infty}^{+\infty}v(\nu)\exp\left[-2\pi\mathrm{i}\nu t\right]\mathrm{d}\nu \tag{3.5}$$

where

$$v(\nu)=\tfrac{1}{2}a(\nu)\exp\left[\mathrm{i}\phi(\nu)\right] \tag{3.6}$$

then the associated analytic signal is

$$V(t)=2\int_0^\infty v(\nu)\exp\left[-2\pi\mathrm{i}\nu t\right]\mathrm{d}\nu. \tag{3.7}$$

Thus $V(t)$ may be derived from the Fourier integral of $V^{(r)}(t)$ by simply suppressing the negative frequencies and multiplying the amplitudes of the positive frequencies by 2. The following relations will be useful later and are simple to prove;

$$\int_{-\infty}^{+\infty} V^*(t)V(t)\,\mathrm{d}t = 2\int_{-\infty}^{+\infty} V^{(r)2}(t)\,\mathrm{d}t = 2\int_{-\infty}^{+\infty} V^{(i)2}(t)\,\mathrm{d}t$$

$$= 2\int_{-\infty}^{+\infty} |v(\nu)|^2\,\mathrm{d}\nu = 4\int_{0}^{\infty} |v(\nu)|^2\,\mathrm{d}\nu. \tag{3.8}$$

Because it is convenient when dealing with stationary random processes we have assumed so far that $V(t)$ is defined for all values of t. In practice, observations will only be carried out over some finite time $-T \leqslant t \leqslant T$, but we are justified in letting $T \to \infty$ if it is long compared with any significant periods characteristic of the light wave (e.g. the coherence time). We must however define the average intensity of the field with respect to a finite time T. Following conventional practice, this may be done as follows. The field is represented by the truncated functions

$$V_T^{(r)}(t) = V^{(r)}(t) \quad \text{when } |t| \leqslant T$$
$$= \quad 0 \quad \text{when } |t| > T \tag{3.9}$$

and by analogy with equation (3.7) we may write

$$V_T(t) = 2\int_0^\infty v_T(\nu)\exp\left(-2\pi i\nu t\right)\mathrm{d}\nu \tag{3.10}$$

and the time average of the intensity† is then given (equation 3.8) by

$$\tfrac{1}{2}\langle V^*(t)V(t)\rangle = \langle V^{(r)2}(t)\rangle = 2\int_0^\infty G(\nu)\,\mathrm{d}\nu \tag{3.11}$$

where

$$G(\nu) = \underset{T\to\infty}{\text{Limit}}\,\frac{|v_T(\nu)|^2}{2T}. \tag{3.12}$$

$G(\nu)\,\mathrm{d}\nu$ is the contribution to the total light intensity made by all the frequency components within the range ν to $\nu+\mathrm{d}\nu$ and is called the *spectral density* of the light; it is loosely called the spectrum of the radiation and is measured by a spectro-radiometer.

3.2 The Interference of Two Partially Coherent Beams of Light

In fig. 3.1 consider the pinholes P_1, P_2 in an opaque screen A illuminated by a source of light S. The pinholes allow light to pass to a second screen B. We shall now discuss the intensity of the light on this second screen.

† Power per unit area normal to the direction of propagation.

D

If the fields at P_1, P_2 are represented by the analytic signals $V_1(t)$, $V_2(t)$, then the field at Q may be written

$$V_Q(t) = k_1 V_1(t) + k_2 V_2(t + \tau) \tag{3.13}$$

where k_1, k_2 are the complex amplitude transmission factors for the paths through the two pinholes and take account of the dimensions of the pinholes, the amplitudes and phases of the diffracted secondary waves and the distances of Q from the pinholes; τ is the difference between the times taken by the radiation from the pinholes to reach Q.

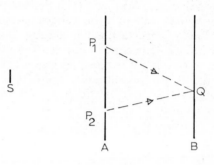

Fig. 3.1.

The intensity at Q is the time-average of the square of the real signal $V_Q^{(r)}(t)$ which, neglecting a constant, we may write

$$I_Q = \langle V_Q^*(t) V_Q(t) \rangle \tag{3.14}$$

where the angle brackets denote the time-average. From equations (3.13) and (3.14)

$$I_Q = |k_1|^2 I_1 + |k_2|^2 I_2 + 2 \operatorname{Re} [k_1^* k_2 \Gamma_{12}(\tau)] \tag{3.15}$$

where $\Gamma_{12}(\tau)$ is called the *mutual coherence function* of the light at the two pinholes and is defined by

$$\Gamma_{12}(\tau) = \langle V_1^*(t) V_2(t + \tau) \rangle. \tag{3.16}$$

We may also write

$$\Gamma_{11}(0) = \langle V_1^*(t) V_1(t) \rangle = I_1$$
$$\Gamma_{22}(0) = \langle V_2^*(t) V_2(t) \rangle = I_2 \tag{3.17}$$

where I_1 and I_2 are the intensities at the two pinholes.

Equation (3.15) can be written in a more general form in terms of the dimensionless *complex degree of coherence* $\gamma_{12}(\tau)$ where

$$\gamma_{12}(\tau) = \{I_1 I_2\}^{-1/2} \Gamma_{12}(\tau) \tag{3.18}$$

34

so that

$$I_Q = I_{1Q} + I_{2Q} + 2\{I_{1Q}I_{2Q}\}^{1/2}\,\mathrm{Re}\,[\gamma_{12}(\tau)] \tag{3.19}$$

where I_{1Q} and I_{2Q} are the intensities produced at Q by the two pinholes separately.

Equation (3.19) shows that the intensity at any point on the second screen can be found from a knowledge of the intensities at that point which would be produced by the pinholes independently and the real part of the complex degree of coherence between the light at the pinholes for the appropriate value of τ. Conversely it is possible, at least in principle, to find the complex degree of coherence of the light at the two pinholes by making measurements of the light on the screen B.

3.3 Spatial Coherence—the Dependence of Coherence on the Angular Size of the Source

Let us suppose that in the interference experiment illustrated in fig. 3.1 the point Q is equidistant from the two pinholes P_1, P_2 so that we may put $\tau = 0$ and write equation (3.19) as

$$I_Q = I_{1Q} + I_{2Q} + 2(I_{1Q}I_{2Q})^{1/2}\,\mathrm{Re}\,(\gamma_{12}(0)). \tag{3.20}$$

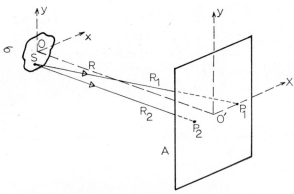

Fig. 3.2. Illustrating the van Cittert–Zernike theorem.

If now we alter the separation between P_1 and P_2, or the angular size of the source, or the wavelength of the light, then the degree of coherence at the pinholes (γ_{12}) will vary also. We shall refer to this as a variation of *spatial coherence* to distinguish it from *temporal coherence* which is discussed in the next section. The dependence of this spatial coherence on these three parameters is the fundamental relationship on which all stellar interferometry depends and it is expressed by the van Cittert–Zernike theorem.

Following Born and Wolf (1959) the van Cittert–Zernike theorem can be presented as follows. Consider the two points P_1, P_2 on the screen A in fig. 3.2 which is illuminated by the extended source σ.

35

For simplicity we shall take the source to lie in a plane parallel to A and to be at a very great distance compared with the size of the source and the separation of the pinholes, so that all the angles between OO' and lines joining points on the source to P_1 and P_2 are extremely small. Furthermore, to separate clearly the effects of spatial and temporal coherence, we shall assume that P_1 and P_2 are equidistant from the source and that the path difference (R_2-R_1) from any point on the source to P_1, P_2 is very small compared with the coherence length $(c/\Delta\nu)$ of the light (see §3.4). Also for simplicity, we shall take the light reaching P_1, P_2 to be quasi-monochromatic; that is to say, the optical bandwidth $\Delta\nu$ is restricted by narrow-band filters (not shown) so that $\Delta\nu/\nu_0 \ll 1$, where ν_0 is the mean frequency of the filter bandwidth. Let us now consider the surface of the source to be divided into a large number of small, independent sources $d\sigma_1$, $d\sigma_2$, . . . with linear dimensions small compared with the mean wavelength λ_0 of the light. Then if $V_{m1}(t)$, $V_{m2}(t)$ are the complex wave amplitudes at P_1, P_2 due to an elementary source $d\sigma_m$, the mutual coherence function of the light at those points is

$$\Gamma_{12}(0) = V_1^*(t)V_2(t) = \sum_m V_{m1}^*(t)V_{m2}(t)$$

$$+ \sum_{m \neq n}\sum V_{m1}^*(t)V_{n2}(t). \tag{3.21}$$

We may put the second term in this expression equal to zero, because there is no correlation between different elements of the source, and write

$$\Gamma_{12}(0) = \sum_m V_{m1}^*(t)V_{m2}(t). \tag{3.22}$$

If $I(S)$ is the intensity per unit area of the source and R_1, R_2 are the distances from a typical point S to P_1, P_2, then it can be shown (Born and Wolf, 1959) that equation (3.22) yields

$$\Gamma_{12}(0) = \int_\sigma (I(S)/R_1 R_2) \exp\left[2\pi i(R_1-R_2)/\lambda_0\right] dS. \tag{3.23}$$

The complex degree of coherence $\gamma_{12}(0)$, as defined by equation (3.18), is therefore

$$\gamma_{12}(0) = (I_1 I_2)^{-1/2}\int_\sigma (I(S)/R_1 R_2) \exp\left[2\pi i(R_1-R_2)/\lambda_0\right] dS \tag{3.24}$$

where

$$I_1 = \int_\sigma (I(S)/R_1^2)\, dS, \quad I_2 = \int_\sigma (I(S)/R_2^2)\, dS. \tag{3.25}$$

Now let x, y, be the coordinates of a point on the source, and X_1, X_2 be the coordinates of the points P_1, P_2 on the screen; we shall take the two sets of axes to be parallel and P_1, P_2 to lie on the X axis. We may then write

$$R_1 - R_2 \approx (X_1{}^2 - X_2{}^2)/2R - (X_1 - X_2)x/R \qquad (3.26)$$

and putting $R_1 \approx R_2 \approx R_3$ we may rewrite equation (3.24) as

$$\gamma_{12}(0) = \frac{\exp(i\psi) \iint\limits_{\sigma} I(x,y) \exp[-2\pi i(X_1 - X_2)x/\lambda_0 R]\, dx\, dy}{\iint\limits_{\sigma} I(x,y)\, dx\, dy} \qquad (3.27)$$

where

$$\psi = (2\pi/\lambda_0)\,[(X_1{}^2 - X_2{}^2)/2R]. \qquad (3.28)$$

Equation (3.27) is a form of the van Cittert–Zernike theorem and is the fundamental relationship we seek. It expresses the fact that the complex degree of coherence $\gamma_{12}(0)$ of the light at the two points P_1, P_2 is given by the normalized Fourier transform of the distribution of intensity over the source, the source being reduced to an equivalent strip distribution parallel to the line joining P_1, P_2. The factor $\exp(i\psi)$ represents the phase-shift $(2\pi/\lambda_0)\,(OP_1 - OP_2)$ and is unity in the case we are considering where P_1, P_2 are equidistant from the source.

In presenting the result in equation (3.27) we have made several assumptions to simplify the discussion. There are two points which are worth noting. First, the assumption that the light reaching P_1, P_2 is quasi-monochromatic is certainly justified in an intensity interfero-meter where, for a number of practical reasons, a narrow-band optical filter must be used. On the other hand, such a filter was not actually used in Michelson's stellar interferometer and in that case it is necessary to convolve the expression for $\gamma_{12}(0)$ with the spectral distribution of the light. A second point is that we have assumed, again for simplicity, that the source lies in a plane and this is clearly not true of a star, even less of a double star. It is however simple to show that our result is valid for a star provided only that its distance is very much greater than any other dimension in the whole system. It should be noted that the distribution of intensity across the source to which we refer is the distribution projected onto a plane normal to the direction of the source from the points P_1, P_2.

3.4 Temporal Coherence—the Dependence of Coherence on Path Difference

We shall now discuss the effect of path difference or relative time delay on the mutual coherence of two beams of light. The question which we seek to answer is illustrated in fig. 3.3.

Suppose a plane wave from a distant source of light S illuminates the two points P_1, P_2 by means of a half-silvered mirror M in such a way that P_1 and P_2 are perfectly coincident when viewed from the source.

If $MP_1 = MP_2$ then the light from S arrives simultaneously at P_1, P_2 and the degree of coherence, if we could measure it, would be $\gamma_{12}(0) = 1$. As we move P_1 to P_1', keeping P_1' and P_2 still coincident as seen from the source, we introduce a time delay $\tau = (MP_1' - MP_1)/c$ between the fields at P_1' and P_2 without altering their space coherence. The problem is to find how $\gamma_{12}(\tau)$ depends upon τ.

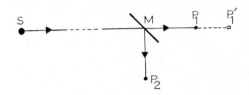

Fig. 3.3.   Illustrating the theory of temporal coherence.

If we represent the waves at P_1, P_2 by the analytic signals $V_1(t)$, $V_2(t+\tau)$ then it can be shown that the mutual coherence function is given by

$$\Gamma_{12}(\tau) = \langle V_1^*(t)V_2(t+\tau)\rangle = 4\int_0^\infty G_{12}(\nu)\exp\left(-2\pi i\nu\tau\right)d\nu \qquad (3.29)$$

and therefore

$$\gamma_{12}(\tau) = \int_0^\infty G_{12}(\nu)\exp\left(-2\pi i\nu\tau\right)d\nu \Big/ \int_0^\infty G_{12}(\nu)\,d\nu \qquad (3.30)$$

where, by analogy with equation (3.12),

$$G_{12}(\nu) = \underset{T\to\infty}{\text{Limit}}\left[\frac{v_{T1}^*(\nu)v_{T2}(\nu)}{2T}\right] \qquad (3.31)$$

and $G_{12}(\nu)$ is called the *mutual spectral density* of the two beams at P_1, P_2.

Equation (3.29) expresses the important result that the mutual coherence varies directly as the Fourier transform of the mutual spectral density of the two beams of light.   If the spectra at P_1, P_2 are identical, as they would be in many practical cases, then $\Gamma_{12}(\tau)$ corresponds to the auto-correlation function of the light.   In that case, equation (3.29) is a statement, in optical terms of the Wiener–Khinchin theorem which is well known in the theory of stationary random processes.   This theorem states that the auto-correlation function of a stationary random process is given by the Fourier transform of its power spectrum, which is the principle of Fourier spectroscopy.

38

If we consider the simple case where the mutual spectral density $G_{12}(\nu)$ is uniform over a narrow bandwidth $\Delta\nu$ about a mean frequency ν_0 and $\Delta\nu/\nu_0 \ll 1$, then from equation (3.30)

$$\gamma_{12}(\tau) = [\sin \pi\nu\tau/\pi\nu\tau] \exp(-2\pi i\nu_0\tau). \qquad (3.32)$$

This function reaches a first zero when the relative time delay is τ_0 or when the path difference is l_0 where

$$\tau_0 = 1/\Delta\nu \quad \text{and} \quad l_0 = c/\Delta\nu. \qquad (3.33)$$

τ_0 is usually referred to as the *coherence time* and l_0 as the *coherence length* of the light.

3.5 *The Correlation between Fluctuations of Intensity*

An intensity interferometer measures the correlation between the *fluctuations of intensity* at two separated points in a partially coherent field. In this section we are concerned to establish only the general principle that such a correlation exists and not to consider any specific method of measurement. Consider again the two points P_1, P_2 illuminated by a distant source of finite angular size (fig. 3.2). Following the discussion given by Mandel (1963) the intensities at P_1, P_2 are

$$I_1(t) = V_1^*(t)V_1(t), \quad I_2(t) = V_2^*(t)V_2(t) \qquad (3.34)$$

and the correlation between these intensities is given by

$$\begin{aligned}
\langle I_1(t)I_2(t+\tau)\rangle &= \langle V_1^*(t)V_1(t)V_2^*(t+\tau)V_2(t+\tau)\rangle \\
&= \langle V_1^{(r)2}(t)V_2^{(r)2}(t+\tau)\rangle + \langle V_1^{(r)2}(t)V_2^{(i)2}(t+\tau)\rangle \\
&\quad + \langle V_1^{(i)2}(t)V_2^{(r)2}(t+\tau)\rangle + \langle V_1^{(i)2}(t)V_2^{(i)2}(t+\tau)\rangle.
\end{aligned} \qquad (3.35)$$

Now $V_1^{(r)}(t)$, $V_2^{(r)}(t)$, $V_1^{(i)}(t)$ and $V_2^{(i)}(t)$ are all Gaussian random variates and we may therefore write

$$\langle V_1^{(r)2}(t)V_2^{(r)2}(t+\tau)\rangle = \tfrac{1}{4}I_1 I_2 + 2[\langle V_1^{(r)}(t)V_2^{(r)}(t+\tau)\rangle]^2 \qquad (3.36)$$

and by expanding this equation it can be shown that

$$\langle V_1^{(r)2}(t)V_2^{(r)2}(t+\tau)\rangle = \tfrac{1}{4}I_1 I_2 + \tfrac{1}{2}\{\mathrm{Re}\,[\Gamma_{12}(\tau)]\}^2. \qquad (3.37)$$

Similarly,

$$\left.\begin{aligned}
\langle V_1^{(i)2}(t)V_2^{(r)2}(t+\tau)\rangle &= \tfrac{1}{4}I_1 I_2 + \tfrac{1}{2}\{\mathrm{Im}\,[\Gamma_{12}(\tau)]\}^2 \\
\langle V_1^{(r)2}(t)V_2^{(i)2}(t+\tau)\rangle &= \tfrac{1}{4}I_1 I_2 + \tfrac{1}{2}\{\mathrm{Im}\,[\Gamma_{12}(\tau)]\}^2 \\
\langle V_1^{(i)2}(t)V_2^{(i)2}(t+\tau)\rangle &= \tfrac{1}{4}I_1 I_2 + \tfrac{1}{2}\{\mathrm{Re}\,[\Gamma_{12}(\tau)]\}^2
\end{aligned}\right\} \qquad (3.38)$$

and substituting these results in equation (3.35) we get

$$\langle I_1(t)I_2(t+\tau)\rangle = I_1 I_2 + \Gamma_{12}^2(\tau) = I_1 I_2[1 + |\gamma_{12}(\tau)|^2]. \qquad (3.39)$$

39

Since we are interested in the *fluctuations* of intensity ΔI_1, ΔI_2, about the mean values $\bar{I}_1$, $\bar{I}_2$ we write,

$$\langle I_1(t)I_2(t+\tau)\rangle = \bar{I}_1\bar{I}_2 + \langle \Delta I_1(t)\Delta I_2(t+\tau)\rangle \qquad (3.40)$$

and from equation (3.39)

$$\langle \Delta I_1(t)\Delta I_2(t+\tau)\rangle = |\Gamma_{12}(\tau)|^2 \qquad (3.41)$$

and

$$\langle \Delta I_1(t)\Delta I_2(t+\tau)\rangle = \bar{I}_1\bar{I}_2|\gamma_{12}(\tau)|^2. \qquad (3.42)$$

Equation (3.42) establishes the basic principles on which an intensity interferometer depends. If the fields at two separated points are partially coherent, then the fluctuations of intensity at these two points are correlated. This correlation is, as one might expect, proportional to the *square* of the degree of coherence $|\gamma_{12}(\tau)|^2$.

At this point we should note that the previous analysis refers to linearly polarized light and that in later discussions of a practical instrument we shall be dealing with unpolarized light. It can be shown (e.g. Mandel, 1963) that for unpolarized light the correlation is half that expected for polarized light because the orthogonal components of the field are uncorrelated. Thus for unpolarized light equation (3.42) is written

$$\langle \Delta I_1(t)\Delta I_2(t+\tau)\rangle = \tfrac{1}{2}\bar{I}_1\bar{I}_2|\gamma_{12}(\tau)|^2. \qquad (3.43)$$

CHAPTER 4

the principles of three types of interferometer

4.1 *The Principles of Michelson's Interferometer*

4.1.1 *Fringe visibility*

In chapter 2 we reviewed briefly Michelson's stellar interferometer in which the light from two small separated mirrors (M_1, M_2 in fig. 2.1) is superimposed in the focal plane of a telescope and the image of a star is seen to be crossed by alternate bright and dark bands called fringes. In operation the two mirrors are separated, the contrast or visibility of the fringes decreases and they eventually disappear. By measuring the separation of the mirrors at which this disappearance occurs it is possible from equation (2.2) to find the angular diameter of the star.

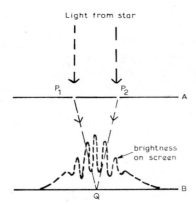

Fig. 4.1. Illustrating the principle of a Michelson interferometer.

We may conveniently discuss the principles of Michelson's interferometer in terms of the simple two-pinhole arrangement shown in fig. 4.1. The two pinholes P_1, P_2 represent the two small mirrors M_1, M_2 and the focal plane is represented by the screen B. The point Q on screen B is equidistant from P_1, P_2. The star is a very distant source of finite angular size which illuminates the pinholes. The 'image' of the star at Q is a circular patch of light which in a practical interferometer corresponds to the diffraction-limited image of a star seen in the small mirrors M_1, M_2. For simplicity we shall restrict the discussion to the case of quasi-monochromatic light with a bandwidth $\Delta\nu$ and mean frequency ν_0 so that $\Delta\nu/\nu_0 \ll 1$. We may now write the complex degree of coherence of the two beams reaching Q as

$$\gamma_{12}(\tau) = |\gamma_{12}(\tau)| \exp i(\alpha - \delta) \qquad (4.1)$$

where α is any initial phase difference between the waves leaving the two pinholes, and δ is the phase difference at frequency ν_0 due to the path difference $(P_1Q - P_2Q)$. From equation (3.19) we may now write

$$I_Q = I_{1Q} + I_{2Q} + (I_{1Q}I_{2Q})^{1/2}|\gamma_{12}(\tau)| \cos{(\alpha - \delta)} \qquad (4.2)$$

where

$$\left. \begin{array}{l} \delta = (2\pi/\lambda_0)\,(P_1Q - P_2Q) \\[2mm] \tau = (P_1Q - P_2Q)/c \end{array} \right\} \qquad (4.3)$$

and

and I_{1Q}, I_{2Q} are the intensities which would be produced at Q by the two pinholes acting independently.

Equation (4.2) shows that the distribution on the screen near Q varies sinusoidally with maxima given by

$$\alpha - \delta = 2m\pi, \quad m = 0,\ \pm 1,\ \pm 2 \ldots \qquad (4.4)$$

This corresponds to a system of bright and dark fringes running across the 'image' of the star in a direction normal to the line joining the pinholes (fig. 2.2). The maxima and minima of these fringes are given by

$$\left. \begin{array}{l} I_{\max} = I_{1Q} + I_{2Q} + 2(I_{1Q}I_{2Q})^{1/2}|\gamma_{12}(\tau)| \\[2mm] I_{\min} = I_{1Q} + I_{2Q} - 2(I_{1Q}I_{2Q})^{1/2}|\gamma_{12}(\tau)| \end{array} \right\} \qquad (4.5)$$

Therefore, if we follow the definition of *fringe visibility* used by Michelson, the fringe visibility V near to Q is given by

$$V_{\text{near Q}} = \frac{I_{\max} - I_{\min}}{I_{\max} + I_{\min}} = \frac{2(I_{1Q}I_{2Q})^{1/2}}{I_{1Q} + I_{2Q}}|\gamma_{12}(\tau)|. \qquad (4.6)$$

In the simple case, where $I_{1Q} = I_{2Q}$ and τ is small $(\tau \ll 1/\Delta\nu)$, we may write

$$V_{\text{near Q}} = |\gamma_{12}(0)|. \qquad (4.7)$$

Thus by measuring the visibility of the fringes it is possible to find the degree of coherence of the light at the two separated mirrors. Furthermore, if these measurements are carried out at several different mirror separations then, as we have already discussed in chapter 3.3, they yield the Fourier transform of the distribution of intensity across the star and hence the angular diameter of the star.

As an example, if the star presents a circular disc of uniform brightness, then the fringe visibility V_d observed with a mirror separation d, is given by

$$V_d = (2J_1(\pi d\theta/\lambda_0))/(\pi d\theta/\lambda_0) \qquad (4.8)$$

where J_1 is a Bessel function of the first order and θ is the angular diameter of the star. This function is illustrated in fig. 2.3 and shows that the first zero, or disappearance of the fringes, occurs when

$$d = 1 \cdot 22\lambda/\theta. \qquad (4.9)$$

Hence, by measuring this critical value of d we can find the star's angular diameter (equivalent uniform disc) from equation (4.9).

It is interesting to note from equation (4.6) that the fringe visibility yields only the modulus of the complex degree of coherence and not the argument. The argument can be found, at least in principle, by measuring the position of the fringes. Thus the fringes near the point Q (fig. 4.1) are displaced in a direction parallel to the line P_1, P_2 by an amount x,

$$x = (\lambda_0/2\pi)(AB/P_1P_2)\alpha \qquad (4.10)$$

where x is measured relative to the position of the fringes which would be produced by light of wavelength λ_0 radiated in phase by P_1, P_2. In practice such a measurement would be difficult, if not impossible, to make with a large interferometer because it requires that the instrument should be pointed at the star with extreme precision; furthermore, atmospheric scintillation introduces random phase-shifts into the light reaching the two small mirrors and causes the fringes to move about rapidly in the focal plane.

4.1.2 *The effects of path differences*

It was pointed out in chapter 2 that a serious disadvantage of Michelson's stellar interferometer is that the two light paths through the instrument must be maintained equal in length with a precision which is hard to achieve in practice. Consider first the effect of small, irregular path differences which are less than the mean wavelength of the light. If the light has a narrow bandwidth, $\Delta\nu/\nu_0 \ll 1$, the effect is to introduce a phase shift ϕ into all frequency components of the light. Equation (4.10) shows that the fringe pattern will then be displaced by an amount proportional to the phase-shift and, when the relative phase at the two small mirrors is changed by π, the bright and dark bands of the fringe pattern will be interchanged; if this happens rapidly compared with the observing time, the fringe visibility will be reduced. Consider, for example, light from a distant star falling on two points P_1, P_2 on the Earth. To reach these points the light must traverse an irregular atmospheric layer which introduces phase shifts

ϕ_1, ϕ_2. If $\gamma_{12}(\tau)$ is the complex degree of coherence between the light at P_1, P_2 which would be observed in the absence of the atmosphere ($\phi_1 = \phi_2 = 0$), then it can be shown (e.g. Beran and Parrent, 1964) that $\gamma_{12}'(\tau)$ in the presence of the atmosphere will be

$$\langle \gamma_{12}'(\tau) \rangle = \gamma_{12}(\tau) \exp i \langle \phi_1 - \phi_2 \rangle \qquad (4.11)$$

where the angle brackets denote time-averages. If now the phase-shifts are random, uncorrelated and distributed with even probability over the range $-\pi$ to $+\pi$, then

$$\langle \phi_1 \rangle = \langle \phi_2 \rangle = \langle \phi_1 \phi_2 \rangle = 0 \qquad (4.12)$$

and hence from equation (4.11)

$$\langle \gamma_{12}'(\tau) \rangle = 0. \qquad (4.13)$$

Thus the time-average of $\gamma_{12}'(\tau)$ tends to zero if it is taken over periods which are comparable with the characteristic period of the fluctuations in phase. It follows, for example, from equation (4.6), that the time-average of the fringe visibility will also be reduced, but the significance of this effect will depend on exactly how the fringes are measured and on the magnitude and frequency spectrum of the phase scintillations introduced by the atmosphere.

We shall now consider the effect of introducing path differences or time delays which are significantly larger than a wavelength or period of the light. From equation (3.32) in the previous discussion of temporal coherence it follows that, if there is a relative time delay of τ between the two beams of light reaching the focus, the coherence factor and therefore also the fringe visibility will be reduced by the factor

$$\sin(\pi \Delta \nu \tau)/(\pi \Delta \nu \tau) \qquad (4.14)$$

where $\Delta \nu$ is the optical bandwidth which we have assumed, for simplicity, to be rectangular. Thus, if we are observing fringes by eye at a mean wavelength of 540 nm, with a bandwidth of say 100 nm, the fringe visibility will decrease with time delay or path difference as shown in fig. 4.2.

From this figure we can see that the fringes will disappear entirely if the path difference is about 3 μm or the time delay is 10^{-14} s. In practice this means that the instrument must be extremely stable mechanically and that it must be pointed at the star with extreme precision. For example, if we require that the fringe visibility is not significantly altered by pointing errors, then the maximum permissible path difference, for a 10 per cent loss in fringe visibility, is about 1 μm, which corresponds to a maximum permissible pointing error of 0·02 seconds of arc for a baseline of 10 m. It is, of course, possible to make these requirements less stringent by reducing the optical bandwidth; for example, if we reduce the bandwidth to 0·1 nm, then for a 10 per cent loss in fringe visibility we can tolerate a path difference of 1 mm or a pointing error of 20 seconds of arc. However, a restriction of the optical bandwidth inevitably reduces the sensitivity of the instrument and it must be remembered that this loss cannot necessarily be compensated by simply increasing the mirror size. The maximum possible size of the mirrors is limited by the effects of atmospheric scintillation and must be less than the characteristic size of the scintillation pattern on the ground. This characteristic size varies with site, weather conditions and zenith angle but is typically about 10 cm.

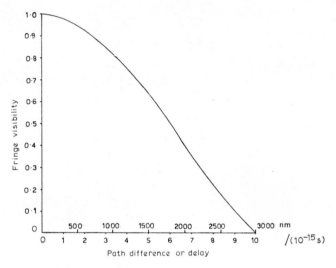

Fig. 4.2. Loss of fringe visibility due to delay or path difference in the arms of a Michelson interferometer ($\lambda = 540$ nm, bandwidth 100 nm).

4.2 The Principles of an Intensity Interferometer using a Linear Multiplier

4.2.1 The fluctuations in the output of a photoelectric detector

In § 3.5 we established the general principle that if the light at two separated points is partially coherent, then the fluctuations of intensity at these two points are also correlated. We shall now enquire how this general principle is applied in an intensity interferometer.

Consider a plane wave of linearly polarized light which illuminates the photoelectric detector in fig. 4.3 and gives rise to an output current $i(t)$. It has been shown by several authors (e.g. Mandel, Sudarshan and Wolf, 1964) that this system may be analysed in terms of a classical electromagnetic wave illuminating a 'quantized' detector and that

45

the probability of emission of a photoelectron $p(t)$ is proportional to the classical measure of the light intensity $I(t)$. Thus we may write

$$p(t)\Delta t = \alpha I(t)\Delta t \qquad (4.15)$$

where α is the quantum efficiency of the detector. It is instructive to note that in this semi-classical picture we do not need to invoke the photon. A discussion of this point has been given by several authors (e.g. Lamb and Scully, 1969).

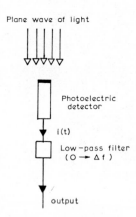

Fig. 4.3. A photoelectric detector illuminated by a plane wave of light.

It follows that the fluctuations in the output of a photoelectric detector may be treated as the fluctuations in the output of a simple square-law detector in which the output current $i(t)$ is proportional to the square of the electric vector of the input wave, so that

$$i(t) = \alpha e I(t) \qquad (4.16)$$

where e is the charge on the electron. If the incident light has a uniform spectral density $G(\nu)$ over a narrow band $\Delta\nu$ (fig. 4.4 a), then we may use the conventional theory of the response of a square law detector to a random noise input (e.g. Rice, 1944) to show that the spectral density of the fluctuations in $i(t)$ is as illustrated in fig. 4.4 b. The low-frequency part of this spectrum extends from zero to $\Delta\nu$ and represents the *difference frequencies* between Fourier components of the light wave. The high-frequency part is centred on $2\nu_0$ and represents the *sum frequencies*; in the present discussion these high frequencies have no significance. Although in principle the low-frequency spectrum extends to extremely high frequencies, and there is experimental evidence that frequencies as high as 10^{10} Hz are preserved in the photoelectric current (Forrester, Gudmundsen and Johnson, 1965), we are concerned here with the practical case where the electrical

46

bandwidth is limited, by the response time of the photoelectric detector and by the bandwidth of the circuits which follow it, to frequencies which are very much less than $\Delta\nu$ and are of the order 10^8 Hz. In fig. 4.5 these limitations (see discussion in § 4.2.2) are represented by a low-pass filter which, for simplicity, has unity gain and a rectangular bandpass from 0 to Δf where $\Delta f \ll \Delta\nu$.

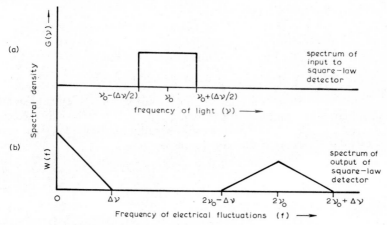

Fig. 4.4. The spectrum of the electrical fluctuations in the output of a square-law detector exposed to light.

Since $\Delta f \ll \Delta\nu$ the spectral density $W(f)$ of the 'rectified fluctuations' in the light is roughly uniform at the output of the filter and it can be shown (e.g. Rice, 1944) that

$$W(f) = 2i_{\mathrm{DC}}^2/\Delta\nu \qquad (4.17)$$

where i_{DC} is the direct current component due to the light. We shall call the fluctuations *wave noise* and at the output of the filter their mean square value is

$$\overline{j_c^2} = W(f)\Delta f = 2i_{\mathrm{DC}}^2 \Delta f/\Delta\nu. \qquad (4.18)$$

In addition to this wave noise there will be the classical *shot noise* j_n^2 due to the finite charge on the electron and the stochastic association between photoemission and the incident light where

$$\overline{j_n^2} = 2ei_{\mathrm{DC}}\Delta f. \qquad (4.19)$$

A more detailed analysis (Hanbury Brown and Twiss, 1957 b) confirms that we simply add the shot *and* wave noise to find the total noise $\overline{j^2}$ at the output of the low-pass filter, so that

$$\overline{j^2} = \overline{j_n^2} + \overline{j_c^2} = 2ei_{\mathrm{DC}}\Delta f + 2i_{\mathrm{DC}}^2 \Delta f/\Delta\nu = 2ei_{\mathrm{DC}}\Delta f[1 + r] \qquad (4.20)$$

where $r = i_{\mathrm{DC}}/e\Delta\nu$ and is the number of photoelectrons per unit time in unit optical bandwidth.

47

To summarize, equation (4.20) shows that the fluctuations in output current from a photoelectric detector, exposed to a plane wave of light, are greater than classical shot noise. The additional component is due to the fluctuations in the incident light wave and we call it *wave noise*.

4.2.2 The correlation between the fluctuations in the output currents of two photoelectric detectors

In § 3.5 we found that the fluctuations in the intensity of the light at two separated points are correlated when these points are illuminated by partially coherent light, and for unpolarized light (equation (3.43)),

$$\langle \Delta I_1(t)\Delta I_2(t+\tau)\rangle = \tfrac{1}{2}I_1 I_2|\gamma_{12}(\tau)|^2. \tag{4.21}$$

Since the wave noise in the output of a photoelectric detector represents the 'rectified fluctuations' in the incident light, it is to be expected that the wave noises from two detectors, exposed to partially coherent light, are correlated, whereas the shot noises are not.

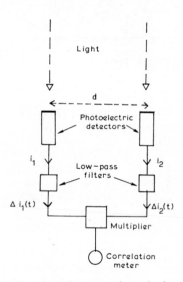

Fig. 4.5. Outline of an elementary intensity interferometer.

Consider the elementary intensity interferometer shown in fig. 4.5, which consists of two photoelectric detectors, followed by identical low-pass filters feeding a linear multiplier. We wish to know how the averaged output of this multiplier depends upon the coherence of the light. We must first remember that in equation (3.43) the answer to this question is given in terms of the correlation between the instantaneous fluctuations of intensity, whereas a practical detector has a finite response time. The process of photoelectric emission must, in our semi-classical picture, be regarded as taking an average of the intensity of the light wave over a few cycles (e.g. Mandel, Sudarshan and Wolf, 1964). Furthermore, there is inevitably some finite response time due to capacity or dispersion of transit times in the detector, and this will clearly average over many cycles of the light. Typically the response time of a complete photomultiplier lies in the range 0·5–5 ns and so, in effect, the output is an average of the light intensity over roughly one million cycles. Mandel (1963) treats this averaging process by writing the current $i(t)$ as

$$i(t) = (\alpha e/T) \int_{-T/2}^{T/2} I(t+t')\, dt' \tag{4.22}$$

where T is the time over which the intensity is averaged. He then shows that for unpolarized light the correlation between the fluctuations in $i(t)$ is

$$\langle \Delta i_1(t)\Delta i_2(t)\rangle = (\alpha^2 e^2 \bar{I}_1 \bar{I}_2/2T^2) \int\!\!\int_{-T/2}^{T/2} |\gamma_{12}(t'-t'')|^2\, dt'\, dt''. \tag{4.23}$$

In the present case, where the spectrum of the light is identical in the two detectors, we may write

$$\gamma_{12}(\tau) = \gamma_{12}(0)\gamma_{11}(\tau) \tag{4.24}$$

and assuming that $T \gg 1/\Delta\nu$ equation (4.23) may be written

$$\langle \Delta i_1(t)\Delta i_2(t)\rangle = \tfrac{1}{2}\alpha^2 e^2 \bar{I}_1 \bar{I}_2 |\gamma_{12}(0)|^2 (\tau_0/T) \tag{4.25}$$

where

$$\tau_0 = \int_{-\infty}^{\infty} |\gamma_{11}(\tau)|^2\, dt. \tag{4.26}$$

Thus the effect of averaging over a resolving time T is to reduce the the correlation by a factor τ_0/T where τ_0 is the coherence length of the light. It is simple to show analytically (e.g. Mandel, 1963) that we may represent this resolving time T by a low-pass filter of bandwidth $\Delta f(\approx 1/T)$ and from equation (3.33) it follows that we may express the coherence time τ_0 in terms of the optical bandwidth $\Delta\nu(\approx 1/\tau_0)$. For the case of simple rectangular bandwidths it can be shown that $\tau_0/T = 2\Delta f/\Delta\nu$ and therefore we may write equation (4.25) as,

$$\langle \Delta i_1(t)\Delta i_2(t)\rangle = \alpha^2 e^2 \bar{I}_1 \bar{I}_2 |\gamma_{12}(0)|^2 (\Delta f/\Delta\nu). \tag{4.27}$$

It follows from this discussion that we may represent the combined response time of the photocathode, photomultiplier and the circuits between the output of the photomultiplier and the linear multiplier by the simple low-pass filter shown in fig. 4.5. In this diagram the photoelectric detectors and the multiplier have zero response time and the low-pass filters have unity gain and zero phase-shift over a bandwidth from 0 to Δf.

49

E

We may now rewrite equation (4.27) as

$$\overline{c(d)} = \langle \Delta i_1(t)\Delta i_2(t)\rangle = e^2 A^2\alpha^2 n^2|\gamma_d(0)|^2\Delta\nu\Delta f \qquad (4.28)$$

where $\overline{c(d)}$ is the correlation or time-average of the multiplier output when the detectors are spaced by a distance d; A is the area of each detector; n is the intensity of the light in photons per unit optical bandwidth per unit area and unit time; and $|\gamma_d(0)|$ is the degree of coherence corresponding to $\tau = 0$ and a baseline d.

From equation (4.28) we see that the correlation observed at any given baseline (d) is proportional to $|\gamma_d(0)|^2$ the *square* of the modulus of the complex degree of coherence. From equation (4.6) it is therefore also proportional to the *square of the fringe visibility* in Michelson's stellar interferometer. The theoretical variation of correlation $c(d)$ with baseline (mirror separation) is shown as a broken line in fig. 2.3 for the case of a star with a circular disc of uniform brightness.

4.2.3 *The signal-to-noise ratio*

The precision with which the correlation $c(d)$ can be measured is limited by fluctuations or noise in the multiplier output. This noise is due principally to statistical fluctuations in the product of the shot noises in the two channels. As we shall see in §4.3.1, under practical conditions wave noise is small compared with shot noise. In the simple case where the two detectors are identical we may take the shot noises in the two channels to be equal and it can be shown (e.g. Hanbury Brown and Twiss, 1957 b) that the r.m.s. noise N in the multiplier output is

$$N(T_0) = \sqrt{2}e^2 A\alpha n\Delta\nu(\Delta f/T_0)^{1/2} \qquad (4.29)$$

where T_0 is the interval over which the multiplier output is averaged. In the same interval the correlation is $c(d)T_0$, so that from equations (4.28) and (4.29) the r.m.s. signal/noise ratio is

$$(S/N)_{\mathrm{RMS}} = c(d)T_0/N(T_0) = A\alpha n|\gamma_d(0)|^2(\Delta f T_0/2)^{1/2}. \qquad (4.30)$$

Equation (4.30) shows that the signal/noise ratio is directly proportional to the light-collecting area A of the detectors and to the quantum efficiency α, and that it is proportional to the square root of the electrical bandwidth $(\Delta f)^{1/2}$ and observing time $T_0^{1/2}$. It is interesting to note that it is independent of the optical bandwidth $\Delta\nu$ but is directly proportional to n the number of photons incident on unit area in unit time and in unit optical bandwidth; it should be noted that this latter quantity is a property of the source and not of the equipment.

We may use equation (4.30) to make a rough estimate of the signal/noise ratio of the Narrabri stellar interferometer. Taking $A = 30$ m², $\alpha = 0\cdot20$, $\lambda = 430$ nm, $\Delta f = 100$ MHz and $T_0 = 1$ h, the signal/noise ratio when observing an unresolved $(\gamma_d(0)=1)$ zero-magnitude star in the zenith $(n \approx 5 \times 10^{-5}$ photons $m^{-2} s^{-1} Hz^{-1})$

$$c(d)/N = 127\Sigma \text{ in 1 hour} \tag{4.31}$$

where Σ represents the sum of a large number of small losses in the equipment (such as loss in the optical system, excess noise in the correlator, etc.) and has a value of about 0·2.

Equation (4.31) illustrates the most serious disadvantage of an intensity interferometer, which is, that it requires very large light collectors, even for bright stars. In fact if we take 3/1 in 1 hour as the lowest workable signal/noise ratio, then we are limited to stars brighter than about magnitude $+2$, despite the fact that the light-collecting area of each detector has the very large value of 30 m².

4.2.4 The effects of path difference or time delay

We have already noted that the principal disadvantage of Michelson's stellar interferometer is that it is seriously affected by very small path differences from the source to the focal plane through the two halves of the instrument. It must therefore be constructed and guided with extreme precision and is seriously affected by atmospheric scintillation. The main reasons for the development and success of the intensity interferometer is that it overcame this problem almost completely.

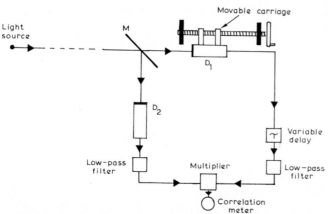

Fig. 4.6.   Intensity interferometer with half-silvered mirror.

Consider the arrangement in fig. 4.6 in which an intensity interfero-meter is used to measure the coherence between two completely coherent beams of light from a half-silvered mirror. A plane wave of quasi-monochromatic light $(\Delta\nu/\nu_0 \ll 1)$ from a distant 'point' source falls on a half-silvered mirror M. It is then directed to the two photoelectric detectors D_1, D_2 which are arranged to be perfectly

51

coincident as seen from the direction of the source. The outputs of these detectors are limited in frequency by two identical low-pass filters with unity gain over the band $O - \Delta f$. The output of these filters is applied to a linear multiplier with unity gain, and the output of the multiplier is averaged and recorded. In one channel there is a variable electrical delay τ; it is also possible to move detector D_1 parallel to the direction of the incident light.

Consider first the case when $MD_1 = MD_2$. The output from the multiplier, the correlation $c(0)$, is then given by equation (4.28) as

$$c(0) = e^2 A^2 \alpha^2 n^2 \Delta \nu \Delta f |\gamma_0(0)|^2. \tag{4.32}$$

If now we move detector D_1 through a very small distance x, such that $x < \lambda$, parallel to the incident light, then a phase-shift $\phi = 2\pi x/\lambda_0$ is introduced into all frequency components of the light at that detector. Under these conditions the complex degree of coherence $\gamma_0(0)$ is modified to $\gamma_0'(0)$ where

$$\gamma_0'(0) \approx \gamma_0(0) \exp{(i\phi)}. \tag{4.33}$$

From equation (4.32) we can now see that a small path-difference or phase-shift has no effect on the measured correlation because the intensity interferometer measures the *modulus* of the complex degree of coherence, and is therefore unaffected by changes in the phase or argument. It follows that random changes of phase, due for example to the atmosphere, have no effect on the measured correlation.

We turn now to the effect of longer path differences or time delays. Consider first the effect of setting $MD_1 = MD_2$ and introducing an electrical delay τ into one channel. This is a classical problem in the theory of random noise and we may use the Wiener–Khinchin theorem, or simply an analogy with equation (3.30), to write

$$c(\tau)/c(0) = \int_0^\infty G_{12}(f) \exp{(-2\pi i f \tau)}\, df / \int_0^\infty G_{12}(f)\, df \tag{4.34}$$

where $G_{12}(f)$ is the mutual spectral density of the *electrical* fluctuations at the outputs of the two filters. In the simple case that we are considering, where the filters are identical and have a uniform band-pass from 0 to Δf, equation (4.34) reduces to

$$c(\tau)/c(0) = \sin{(\pi \Delta f \tau)}/(\pi \Delta f \tau). \tag{4.35}$$

Thus the variation of correlation with time delay is given by the Fourier transform of the spectral distribution of the *electrical* fluctuations applied to the multiplier or, more precisely, by the transform of their mutual or cross-spectral density.

An alternative way of introducing a time delay τ is to move one detector through a distance x parallel to the incident light. Such a path difference is equivalent to a time delay, provided two conditions are satisfied. A first condition is that there must be no significant dispersion of velocity in the medium over a light bandwidth equal to the highest electrical frequency (Δf) passed by the filters. We may express this requirement as

$$x \ll \lambda/(\Delta n \Delta f) \qquad (4.36)$$

where Δn is the change in refractive index of the medium per unit optical bandwidth.

A second condition is that the relative displacement of the detector must not alter the spatial coherence; that is to say, it must not alter the *difference* in path length from points on the source to the detector. It is clear that this condition is satisfied in observation of stars if

$$x \ll \lambda^2/\theta^2 \qquad (4.37)$$

where θ is the apparent angular size of the star.

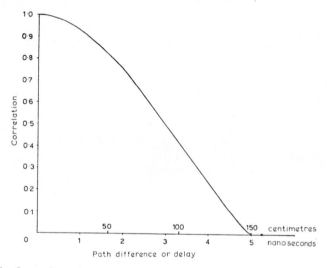

Fig. 4.7.   Loss of correlation due to path difference or differential delay in the arms of an intensity interferometer (bandwidth$=100$ MHz).

Thus, provided the two conditions outlined above are satisfied, we may use equation (4.34) to find the effect of the path differences or electrical delays on the measured correlation. Fig. 4.7 illustrates this effect for the practical case of an intensity interferometer with a rectangular bandwidth of 100 MHz. It shows that there is a 10 per cent loss of correlation for a time delay of or a path difference of about 30 cm. If now we compare these figures with those for a Michelson interferometer then, from figs. 4.2 and 4.7, we see that the intensity interferometer is far less sensitive (roughly 10^5 times) to the effects of path difference or time delays. The practical consequences of this

53

rather surprising result are first, the comparative ease of building and operating a stellar intensity interferometer with the very long baselines (hundreds of metres) which are required to resolve a reasonable sample of the hotter types of stars and secondly, as we shall discuss further in § 5.8, the fact that the measurements are not significantly affected by atmospheric scintillation.

4.3 The Principles of an Intensity Interferometer using a Coincidence Counter

4.3.1 Fluctuations in the output of a photon-counting detector

In this section we shall take a brief look at an alternative form of intensity interferometer based not upon the linear multiplication of the fluctuations in the output currents of two detectors but upon counting coincidences between the arrival times of individual photons. Such an instrument appears to have no practical application to astronomy, nevertheless the discussion of its principles and the experimental tests described later played a useful part in establishing and clarifying the theory of intensity interferometry.

Consider first a single photoelectric detector illuminated by a plane wave of linearly polarized quasi-monochromatic ($\Delta \nu / \nu_0 \ll 1$) light. Let us suppose that the gain of this detector, for example, a photo-multiplier, is such that the pulses in the output current due to single photons can be counted. If the intensity of the light at the photo-cathode, averaged over a few cycles, is $I(t)$, then (see § 4.2.1) the probability that a photoelectron is emitted in a short time dt is $\alpha I(t) dt$, where α is the quantum efficiency assumed to be constant over the optical bandwidth $\Delta \nu$. The mean number of photoelectrons ($\bar{n}$) which will be counted in an interval T is therefore,

$$\bar{n} = \alpha \bar{I} T \tag{4.38}$$

and it can be shown (e.g. Purcell 1956, Mandel, 1963) that the variance $\overline{(\Delta n)^2}$ in this count will be

$$\overline{(\Delta n)^2} = \bar{n}(1 + \bar{n}) \ldots \quad T \ll 1/\Delta \nu \tag{4.39}$$

and

$$\overline{(\Delta n)^2} = \bar{n}(1 + \bar{n}\tau_0/T) \ldots \quad T \gg 1/\Delta \nu \tag{4.40}$$

where

$$\tau_0 = \int\limits_{-\infty}^{\infty} |\gamma_{11}(\tau)|^2 \, d\tau \tag{4.41}$$

and $\gamma_{11}(\tau)$ is the normalized auto-correlation function of the incident light defined by

$$\gamma_{11}(\tau) = \int\limits_{0}^{\infty} G(\nu) \exp(-2\pi i \nu \tau) \, d\nu \Big/ \int\limits_{0}^{\infty} G(\nu) \, d\nu \tag{4.42}$$

54

and $G(\nu)$ is its spectral density. In this case we may identify τ_0 with the coherence time of the light (see equation (3.33)) and, if the spectral density is uniform over the optical bandwidth $\Delta\nu$, write

$$\overline{(\Delta n)^2} = \bar{n}(1 + \bar{n}/\Delta\nu T). \tag{4.43}$$

Equations (4.39) and (4.40) express the fact, noted in §4.2.1, that the fluctuations in the output current of a photoelectric detector exposed to a plane wave of light are greater than the expected value $(\overline{\Delta n^2} = \bar{n})$ in a simple random stream. This excess noise is usually very small; for example, if the source of light produces 10^6 photo-electrons per second and has a bandwidth of 10 nm at 500 nm, then, in a resolving time of 1 ns, the excess noise $(\bar{n}/\Delta\nu T)$ increases the simple random noise $(\bar{n})$ by 1 part in 10^7. For this reason the excess noise in the output of a photoelectric detector is hard to measure directly.

In the previous discussion in §4.2.1 we identified this excess noise as *wave noise* due to fluctuations in the intensity of the incident light, but in the present context, where we are concerned with counting discrete photoelectrons, it is more appropriate to use the 'particle' picture and regard the additional noise as being due to the 'bunching' of photons which obey Bose–Einstein statistics. Intuitively one can see that this bunching must increase the classical shot noise which is due to the purely stochastic association between the emission of a single photoelectron and the light intensity. Put simply, the tendency of photons to bunch, following Bose–Einstein statistics, increases the output fluctuations.

In this connection it is instructive to evaluate the probability of obtaining two counts separated by a time interval τ. Mandel (1963) shows that, for a beam of polarized light, the conditional probability $p_c(\tau)\,d\tau$ of obtaining a second count τ sec after a first is

$$p_c(\tau)\,d\tau = \alpha\bar{I}[1 + |\gamma_{11}(\tau)|^2]\,d\tau. \tag{4.44}$$

Because

$$|\gamma_{11}(\tau)| \approx 1 \quad \text{when} \quad \tau \ll 1/\Delta\nu \tag{4.45}$$

and

$$|\gamma_{11}(\tau)| \approx 0 \quad \text{when} \quad \tau \gg 1/\Delta\nu \tag{4.46}$$

it follows from equation (4.44) that

$$p_c(\tau)\,d\tau \approx 2\alpha\bar{I}\,d\tau \quad \text{when} \quad \tau \ll 1/\Delta\nu \tag{4.47}$$

and

$$p_c(\tau)\,d\tau \approx \alpha\bar{I}\,d\tau \quad \text{when} \quad \tau \gg 1/\Delta\nu \tag{4.48}$$

Thus, when the resolving time is very short $(\tau \ll 1/\Delta\nu)$, the conditional probability of detecting a second photoelectron increases to twice the value $(\alpha\bar{I}\,d\tau)$ expected for a simple random stream of independent

55

particles; on the other hand, when the resolving time is long ($\tau \gg 1/\Delta\nu$), the conditional probability is simply equal to that for the simple random stream. The variation of this conditional probability with the time interval τ is illustrated in fig. 4.8 for a plane polarized wave of light with a Gaussian spectral profile of width $\Delta\nu$. The conditional probability has been normalized to that for a stream of independent particles. Mandel (1963) points out that this bunching, also by a factor of 2, is exhibited by the density fluctuations in any boson gas.

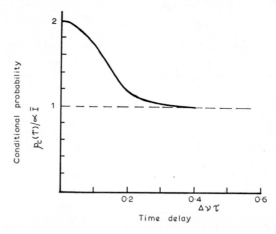

Fig. 4.8.   The conditional probability $p_c(\tau)\mathrm{d}\tau$ of detecting a second photoelectron after an interval τ in a plane wave of linearly polarized light with a Gaussian spectral profile of width $\Delta\nu$. The probability is normalized by the expected value $(\alpha \bar{I}\,\mathrm{d}\tau)$ for a random stream of independent particles.   From Mandel (1963)·

4.3.2   *Coincidence-counting*

Consider the arrangement shown in fig. 4.9. The two photo-multipliers, P_1, P_2 are separated by a distance d and illuminated by unpolarized quasi-monochromatic light ($\Delta\nu/\nu_0 \ll 1$). The two counters N_1, N_2 register the number of photoelectrons in each channel and the counter N_c registers a coincidence when two pulses arrive within a time τ_c. The average counting rates are

$$\bar{N}_1 = \alpha_1 I_1, \quad \bar{N}_2 = \alpha_2 I_2 \tag{4.49}$$

It can be shown (e.g. Mandel, 1963, Hanbury Brown and Twiss, 1957 c) that the coincidence rate $\bar{N}_c$ is

$$\bar{N}_c = \bar{N}_1 \bar{N}_2 [2\tau_c + \tfrac{1}{2}|\gamma_d(0)|^2 \int_{-\tau_c}^{\tau_c} |\gamma_{11}(\tau)|^2 \,\mathrm{d}\tau] \tag{4.50}$$

where it has been assumed for simplicity that the spectral distribution of the light at the two detectors is identical and that their quantum

56

efficiencies are constant over the optical bandwidth. If the resolving time τ_c is either much longer than the coherence time of the light $(1/\Delta\nu)$ or much shorter, equation (4.50) can be simplified as follows: when

$$2\tau_c \gg 1/\Delta\nu$$

$$\bar{N}_c = \bar{N}_1\bar{N}_2 2\tau_c[1 + \tfrac{1}{2}|\gamma_d(0)|^2\tau_0/2\tau_c] \qquad (4.51)$$

and when

$$2\tau_c \ll 1/\Delta\nu$$

$$\bar{N}_c = \bar{N}_1\bar{N}_2 2\tau_c[1 + \tfrac{1}{2}|\gamma_d(0)|^2] \qquad (4.52)$$

where $\tau_0 = 1/\Delta\nu$ as defined previously in equation (4.41).

Equations (4.51) and (4.52) show that, when there is no coherence between the light at the two detectors $(|\gamma_d| = 0)$, the coincidence rate is $\bar{N}_1\bar{N}_2 2\tau_c$, which is the expected rate for two uncorrelated streams of photoelectrons. However, when the light is coherent $(|\gamma_d| > 0)$, there is an excess coincidence rate which depends directly on the square of the degree of coherence. It is therefore possible, at least in principle, to measure the angular distribution of intensity across a source by observing the excess coincidence rate as a function of the spacing between the detectors.

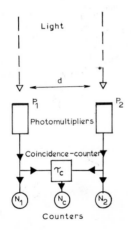

Fig. 4.9. A coincidence-counting intensity interferometer.

It is interesting to note that when the resolving time is short $(2\tau_c \ll 1/\Delta\nu)$ the coincidence rate is increased by $\tfrac{3}{2}$ for unpolarized light; for linearly polarized light this factor would be 2, as one would expect from fig. 4.8. In the more practical case where the resolving time is long compared with the coherence time $(2\tau_c \gg 1/\Delta\nu)$, the excess coincidences are simply reduced by the factor $\tau_0/2\tau_c$.

4.3.3 *The signal-to-noise ratio*

Taking the practical case appropriate to the measurement of stars, where the resolving time is long compared with the coherence time $(2\tau_c \gg 1/\Delta\nu)$, the r.m.s. signal/noise ratio is given by the ratio of the number of excess coincidences in an observing time T_0 to the r.m.s. fluctuations in the random coincidence rate.

The number of random coincidences in a time T_0 is $c_R(T_0)$, where

$$c_R(T_0) = N_1 N_2 2\tau_c T_0. \tag{4.53}$$

Hence, from equations (4.51) and (4.53), it can be shown that the signal/noise ratio is

$$(S/N)_{RMS} = \tfrac{1}{2}(N_1 N_2)^{1/2}\tau_0(T_0/2\tau_c)^{1/2}|\gamma_d(0)|^2. \tag{4.54}$$

A comparison of equations (4.30) and (4.54) shows that the 'linear multiplier' and the 'coincidence counter' give about the same signal/noise ratio when they receive the same number of photoelectrons per second per unit optical bandwidth ($A\alpha n$), provided that the resolving time τ_0 of the coincidence circuit is roughly equal to the reciprocal electrical bandwidth ($1/\Delta f$) of the linear-multiplier circuit. However, it is very difficult in practice to apply the 'coincidence-counting' interferometer to the measurement of stars although, as we shall see in § 6.2, it can be made to work in the laboratory. The difficulty is due to the finite resolving times of practical detectors and counters which severely limit the number of photoelectrons per second that can be counted separately. As an example, consider again the use of an interferometer to observe an unresolved zero-magnitude star in the zenith where $n = 5 \times 10^{-5}$ photons s^{-1} m^{-2} Hz^{-1} and $\gamma_d(0) = 1$. In § 4.2.3 it was shown that the signal/noise ratio in a 'linear-multiplier' system would be 25/1 in one hour's observation. In principle we can achieve the same result with the coincidence-counter by making the resolving time $\tau_c \approx 10$ ns but, unless we greatly restrict the optical bandwidth, the primary photoemission current will be so great that it will be impossible to count individual photoelectrons. Thus, if the maximum possible counting rate is 10^7 photoelectrons per second, the optical bandwidth cannot exceed 3×10^{10} Hz or about 0·02 nm. Although such a narrow bandwidth can be achieved by rather elaborate filters it would be immensely difficult to apply them to a stellar interferometer. The necessary filters would introduce an unacceptable loss of light in the pass-band; they would be difficult to match with the required precision; and they would demand a light beam with better collimation than can be achieved with very large and crude reflectors. In practice this means that coincidence-counting interferometers can only be applied in the laboratory to observations of intense sources of light with narrow bandwidths and they have no apparent application to astronomy.

CHAPTER 5

the theory of a practical stellar intensity interferometer

5.1 Introduction

In the preceding chapters we have reviewed the general principles of an intensity interferometer and derived formulae for the correlation, noise, etc. However, the assumptions underlying these formulae are too idealized for them to apply to a practical instrument. In practice it is necessary to take account of a variety of factors; for example, the apertures of the light detectors may have a finite size, the quantum efficiency and spectral response may be different at the two photo-cathodes, there may be a loss of light in the optical system and the light may be partially polarized; furthermore, the electrical frequency response of the photomultipliers and amplifiers may not be identical in the two channels, there may be excess noise introduced by the electronics and there will inevitably be some loss of correlation in the whole system. In this chapter we shall set out working formulae from which the performance of practical interferometers has been calculated.

5.2 Correlation between Small Apertures

It has been shown (Hanbury Brown and Twiss, 1957 c) that, in the case where the apertures of the light detectors are small, we must rewrite the simple expression for correlation $c(d)$ in equation (4.28) as

$$\overline{c(d)} = e^2 A_1 A_2 \beta_0 \overline{\Gamma^2(d)} \alpha^2(\nu_0) g^2(\nu_0) n^2(\nu_0) \sigma B_0 \epsilon b_{\mathrm{v}} |F_{\mathrm{max}}|^2 \qquad (5.1)$$

where A_1, A_2 are the areas of the two light detectors; $(1-\epsilon)$ is the fraction of the correlation lost in the electronics; B_0 is the effective bandwidth of the light defined by

$$B_0 = [\int_0^\infty \alpha_1(\nu) g_1(\nu) n_1(\nu)\, d\nu \int_0^\infty \alpha_2(\nu) g_2(\nu) n_2(\nu)\, d\nu]^{1/2} / \alpha(\nu_0) g(\nu_0) n(\nu_0) \qquad (5.2)$$

where

$$n_1(\nu) = n_{1a}(\nu) + n_{1b}(\nu) \qquad (5.3)$$

$$n_2(\nu) = n_{2a}(\nu) + n_{2b}(\nu) \qquad (5.4)$$

$$\alpha^2(\nu_0) g^2(\nu_0) n^2(\nu_0) = \alpha_1(\nu_0) \alpha_2(\nu_0) g_1(\nu_0) g_2(\nu_0) n_1(\nu_0) n_2(\nu_0) \qquad (5.5)$$

and $n(\nu)$ is the number of quanta incident in unit time and bandwidth on unit area of the detectors; $g(\nu)$ is the transmittance of the optical system at a frequency ν; $\alpha(\nu)$ is the quantum efficiency at a frequency ν and ν_0 is the mid-band frequency of the light reaching the photo-cathodes; the subscripts a, b refer to two orthogonal directions of polarization; σ is a normalized spectral density function and is defined by

$$\sigma = \int_0^\infty \alpha_1(\nu)g_1(\nu)n_1(\nu)\alpha_2(\nu)g_2(\nu)n_2(\nu)\mathrm{d}\nu / B_0\alpha^2(\nu_0)g^2(\nu_0)n^2(\nu_0) \quad (5.6)$$

and the polarization factor β_0 is defined by

$$\beta_0 = 2[n_{1a}(\nu)n_{2a}(\nu) + n_{1b}(\nu)n_{2b}(\nu)]/n_1(\nu)n_2(\nu) \quad (5.7)$$

so that $\beta_0 = 1$ when $n_a(\nu) = n_b(\nu)$ as in the case of randomly polarized light. The symbol $\overline{\Gamma^2(d)}$ is called the normalized *correlation factor* and is defined by

$$\overline{\Gamma^2(d)} = \overline{c(d)}/\overline{c(0)} \quad (5.8)$$

where $c(d)$ and $c(0)$ are the correlations with baselines d and zero respectively under identical conditions of light flux and observing time and where

$$\overline{\Gamma^2(d)} = \frac{\int_0^\infty |\gamma_d(\nu)|^2\alpha_1(\nu)g_1(\nu)\alpha_2(\nu)g_2(\nu)[n_{1a}(\nu)n_{2a}(\nu) + n_{1b}(\nu)n_{2b}(\nu)]\,\mathrm{d}\nu}{\int_0^\infty \alpha_1(\nu)g_1(\nu)\alpha_2(\nu)g_2(\nu)[n_{1a}(\nu)n_{2a}(\nu) + n_{1b}(\nu)n_{2b}(\nu)]\,\mathrm{d}\nu}$$

$$(5.9)$$

and $|\gamma_d(\nu)|$ is the degree of coherence at baseline d and frequency ν. The effective cross-correlation electrical bandwidth of the correlator b_v is defined by

$$|F_{\max}|^2 b_\mathrm{v} = \tfrac{1}{2}\int_0^\infty [F_1(f)F_2^*(f) + F_1^*(f)F_2(f)]\,\mathrm{d}f \quad (5.10)$$

where $|F_{\max}|$ is the maximum value of $\tfrac{1}{2}[F_1(f)F_2^*(f) + F_1^*(f)F_2(f)]$ and $F(f)$ is the gain of each channel at a frequency f and includes the photomultipliers.

5.3   *Correlation between Large Apertures*

When the aperture of the light detectors is a significant fraction of the baseline necessary to resolve the source, then the light is no longer fully coherent across the individual apertures. This problem has been discussed by Hanbury Brown and Twiss (1957 c). They show that the correlation is reduced by the *partial coherence factor* Δ and that $\overline{\Gamma^2(d)}$ in equation (5.1) can be written $\overline{\Delta\Gamma^2(d)}$.

The general expression for $\overline{\Delta}$ is very cumbersome and is not worth reproducing here; it depends upon the angular size and shape of the source, the size and shape of the detectors and on the spectral distribution of the light. However, we can simplify the discussion by restricting it to the practical case where the bandwidth of the light is so narrow that both $\Gamma^2(d)$ and Δ may be taken as constant over the light bandwidth and we may write

$$\overline{\Delta\Gamma^2(d)} = \Delta(\nu_0)\Gamma^2(\nu_0, d). \tag{5.11}$$

In practice $\Delta(\nu_0)\Gamma^2(\nu_0, d)$ must be evaluated for each specific case by means of a computer. The procedure is to consider the detector to be divided into a number of small elementary areas. The correlation between an elementary area A on one detector and another elementary area B on the other detector is then calculated for a given value of d. This calculation takes into account, of course, the length of the baseline AB and the position angle of AB with respect to the source. The calculation is then repeated for every possible pair of points on the two detectors. Finally the total correlation is found for all possible pairs of points and, after normalizing, yields $\Delta(\nu_0)\Gamma^2(\nu_0, d)$.

As an example, if (x_1, y_1) and (x_2, y_2) represent points on two detectors whose centres are separated by a distance d, then for a source with a uniform circular disc of angular diameter θ

$$\Delta(\nu_0)\Gamma^2(\nu_0, d) = \frac{1}{A_1 A_2} \iiiint \left[\frac{2J_1(\xi)}{\xi}\right]^2 dx_1\, dx_2\, dy_1\, dy_2 \tag{5.12}$$

where

$$\xi = (\pi\theta\nu_0/c)[(x_1 - x_2)^2 + (y_1 - y_2)^2]^{1/2} \tag{5.13}$$

and the integral is taken over the areas A_1, A_2 of the detectors. Fig. 5.1 (a) shows how the partial coherence factor Δ varies with the size of the detectors in the case where a circular source of uniform surface brightness is viewed by two identical circular detectors with a diameter D. Fig. 5.1 (b) shows the normalized correlation factor $\Gamma^2(\nu_0, d)$ as a function of baseline length d for the same configuration; the shape of this curve depends slightly on Δ and is shown for two cases, small apertures $\Delta(\nu_0) = 0.99$ and large apertures $\Delta(\nu_0) = 0.90$.

To summarize, in the general case where the aperture of the detector is large enough to resolve the star partially, the correlation is given by

$$c(d) = e^2 A_1 A_2 \beta_0 \Delta(\nu_0)\Gamma^2(\nu_0, d)\alpha^2(\nu_0)g^2(\nu_0)n^2(\nu_0)B_0\epsilon b_v |F_{\max}|^2 \tag{5.14}$$

where $\Delta(\nu_0)$ the partial coherence factor and $\Gamma^2(\nu_0, d)$ the normalized correlation factor must both be evaluated for the particular size and shape of the source and light detectors. In the simpler case where the aperture of the detectors is small $(\Delta(\nu_0) = 1)$, $\Gamma^2(\nu_0, d)$ is simply the square of the degree of coherence and, as we have seen in §3.3, it is therefore proportional to the square of the modulus of the normalized Fourier transform of the angular distribution of intensity across the equivalent line source.

61

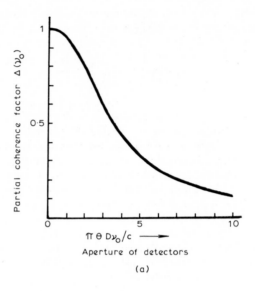

(a)

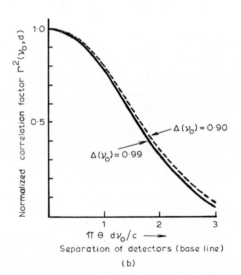

(b)

Fig. 5.1. (*a*) The variation of the partial coherence factor $\Delta(\nu_0)$ with the aperture of the detectors, calculated for a circular source of angular diameter θ viewed by two identical circular apertures with diameter D. (*b*) The variation of the normalized correlation factor $\Gamma^2(\nu_0, d)$ with detector separation, calculated for a circular source of angular diameter θ viewed by two identical apertures with $\Delta(\nu_0)=0\cdot99$ (————), and $\Delta(\nu_0)=0\cdot90$ (– – – –). From Hanbury Brown and Twiss (1958 a).

5.4 The Noise

To calculate the r.m.s. uncertainty $N(T_0)$, or noise, in the correlation measured over an observing time T_0, it is necessary to rewrite equation (4.29) as,

$$N(T_0) = e^2 (A_1 A_2)^{1/2} \alpha(\nu_0) g(\nu_0) n(\nu_0) B_0 (\mu/(\mu-1))(1+\delta)(1+a)$$
$$(2 b_v \eta/T_0)^{1/2} |F_{\max}|^2 \qquad (5.15)$$

where $\mu/(\mu-1)$ represents the excess noise introduced by the photomultiplier chain, and μ is approximately equal to the multiplication factor of the first stage; $(1+\delta)$ allows for excess noise introduced by the correlator and is the ratio of the actual noise at the output of the correlator to the noise due to the outputs of the photomultipliers; $(1+a)$ allows for the presence of stray light and dark current, and is the ratio of the total anode currents of the photomultipliers to the anode currents due to light from the source; η is the normalized spectral density of the cross-correlation frequency response of the two channels of the correlator including the frequency response of the photomultipliers, and is defined by

$$\eta = \int_0^\infty |F_1{}^2(f) F_2{}^2(f)| \, \mathrm{d}f / b_v |F_{\max}|^4. \qquad (5.16)$$

5.5 The Signal-to-Noise Ratio

If we compare equations (5.14) and (5.15) we see that both the correlation $c(d)$ and the noise $N(T_0)$ are linearly proportional to $|F_{\max}|^2$ and their ratio is independent of the gain of the equipment. It is therefore convenient to work in terms of the signal/noise ratio (S/N). From equations (5.14) and (5.15) the r.m.s. signal/noise ratio, for the general case where the apertures of the light detectors are large and the correlation is integrated for a time T_0, is given by

$$(S/N)_{\mathrm{RMS}} = \overline{c(d)} T_0 / N(T_0)$$

$$= (A_1 A_2)^{1/2} \alpha(\nu_0) g(\nu_0) n(\nu_0) \Delta(\nu_0) \Gamma^2(\nu_0, d) \epsilon \beta_0 \sigma((\mu-1)/\mu)$$
$$(b_v T_0/2\eta)^{1/2}/(1+a)(1+\delta) \qquad (5.17)$$

5.6 The Maximum Possible Signal-to-Noise Ratio

Equation (5.17) shows that the signal/noise ratio is proportional to $(A_1 A_2)^{1/2}$ where A_1, A_2 are the areas of the light detectors. At first sight it would therefore appear possible to increase the signal/noise ratio on any source by simply increasing the size of the detectors. However, this is not so; as the size of the detectors is increased they start to resolve the source and the partial coherence factor $\Delta(\nu_0)$ decreases. At the limit when the source is completely resolved by the detectors, the signal/noise ratio approaches monotonically a value determined by the effective black-body temperature of the source at the appropriate wavelength and is independent of the source shape. For an interferometer of the type described in § 6.1, in which the light is divided between the two detectors by a half-silvered mirror, it has been shown (Hanbury Brown and Twiss, 1957 c) that the maximum possible signal/noise ratio, in an observation of duration T_0 of a source with a black-body temperature Θ_0 at a frequency ν_0, is

$$(S/N)_{\max} = \frac{[\alpha(\nu_0)\beta_0\sigma\epsilon g(\nu_0)/(1+a)(1+\delta)]((\mu-1)/\mu)(b_v T_0/2\eta)^{1/2}}{\exp\left[h\nu_0/(k\Theta_0-1)\right]}$$

(5.18)

It is interesting to note that, when the source is completely resolved by the detectors and equation (5.18) applies, the signal/noise ratio depends only on the temperature, and the equipment is, in effect, a pyrometer. An important consequence is that there is a lower limit to the temperature of a source which can be measured with an intensity interferometer. As we shall note in § 11.1, this fact limits the range of spectral types of stars which can be usefully observed.

5.7   A Theoretical Estimate of the Effects of Čerenkov Light

Possible sources of correlation when observing a star with an intensity interferometer are the Čerenkov light pulses produced by cosmic rays entering the Earth's atmosphere. Roughly speaking, the primary cosmic ray produces a shower of particles which in turn produce a pool of light with a radius of the order of 100 m and a duration of the order of 10 ns at the Earth's surface. If both reflectors lie within this pool and the light is produced within their field of view, then correlated pulses will reach the multiplier of the correlator. Unless the correlation due to these pulses is small compared with that from the star, it will produce errors in the measurements of angular size. It is therefore important that an estimate of this unwanted correlation should be made and that the results should be confirmed, if possible, by experiment.

An estimate has been published by Hanbury Brown, Davis and Allen (1969). They took $p(n_c).dn_c.dt.d\Omega$ to be the probability that a cosmic-ray event occurs in a time dt such that, in the absence of atmospheric extinction, n_c photons would reach unit area of both reflectors in unit frequency band from a solid angle $d\Omega$ of sky. They further assumed that these photons arrive in a time interval τ_c such that $\tau_c b_v \geqslant 1$ where b_v is the electrical bandwidth of the electronic equipment including the phototubes, and that the response of the electronic correlator is linear to a burst of N photoelectrons, emitted in time τ_c, for $N < N(\max)$ and then saturates and is independent of N.

Under these conditions it can be shown that the time-averaged correlation $\bar{c}_{cv}$ produced by all events is

$$\bar{c}_{cv} = G\Omega_0 \tau_c^{-1} \left[\int_{n_c(\min)}^{n_c(\max)} X_z^2 n_c^2 p(n_c)\,\mathrm{d}n_c + \int_{n_c(\max)}^{\infty} N^2(\max)p(n_c)\,\mathrm{d}n_c \right]$$
(5.19)

where G is a constant which includes the gain of the equipment; Ω_0 is the solid angle of the field of view of each reflector; $n_c(\min)$ is the value of n_c which yields only one photoelectron; $n_c(\max)$ is the value of n_c which yields $N(\max)$ photoelectrons. The quantity $X_z n_c$ is the number of photoelectrons produced in each phototube by a flux of n_c photons arriving from a zenith angle z, and is given by

$$X_z = Af(\Omega_0/\Omega_{cv}) \int_{\Delta_v} \alpha(v)g(v)\rho(v)^{\sec z}\,\mathrm{d}v \qquad (5.20)$$

where A is the area of each reflector; Δv is the total optical bandwidth; $\alpha(v)$ is the quantum efficiency of the phototube; $g(v)$ is the transmission of the optical system; $\rho(v)^{\sec z}$ is the atmospheric transmission at a zenith angle z; the spectrum of the Čerenkov light is taken to be such that $n_c(v)$ is a constant; $f(\Omega_0/\Omega_{cv})$ is the fraction of the total light in the Čerenkov flash which falls within the angular field of view of the reflectors. From the work of Jelly and Galbraith (1955) they took the distribution of light pulses to be

$$p(n_c) = K_0 n_c^{-2.6} \cos^3 z. \qquad (5.21)$$

Substituting for $p(n_c)$ in equation (5.19) and putting $n_c(\min) = X_z^{-1}$ and $n_c(\max) = N(\max)X_z^{-1}$, then by integration,

$$\bar{c}_{cv} = G\Omega_0 \tau_c^{-1} K_0 \cos^3 z X_z^{1.6}[3 \cdot 12N(\max)^{0.4} - 2\cdot 5]. \qquad (5.22)$$

Hanbury Brown et al. used equation (5.22) to estimate the correlation due to Čerenkov light under normal working conditions on a star in the stellar interferometer at Narrabri. They assumed that $N(\max)$, the saturation level of the correlator, is ten times the r.m.s. noise level due to the light from the star and calculated the expected correlation $\bar{c}_{cv}$ for a wavelength of 443 nm with a total bandwidth of 10 nm. They took the value of the constant K_0 from the direct measurements of the Čerenkov pulse rate made at Narrabri (described in §11.9). Their results are shown by the broken lines in fig. 5.2 where $\log \bar{c}_{cv}$ is plotted as a function of the blue magnitude B of a star under observation in the zenith. In their discussion they reached the conclusion that the principal uncertainty in their estimate is in the exponent of the integral pulse height spectrum which they took to be $-1\cdot6$. They therefore repeated the calculations for extreme values of this index of $-1\cdot0$ and $-2\cdot0$ and the results are shown in fig. 5.2. They suggest that the true correlation must be somewhere between these extremes.

65

It should be noted that in making these calculations, Hanbury Brown *et al.* assumed that the angular field of view Ω_0 of the reflectors (diameter $= 6\cdot5$ m) is small compared with Ω_{cv} the angular distribution of the Čerenkov flash; there is both theoretical (Zatsepin, 1964) and experimental evidence (Hill and Porter, 1961) to support this assumption. It should also be noted that their calculations refer to a separation between the reflectors of 10 m; this was chosen because it is the minimum possible separation and it is under these conditions that one expects the correlation due to Čerenkov light to be a maximum.

Finally, the full line in fig. 5.2 shows, on the same scale, the correlation to be expected from an unresolved star of blue magnitude B in the zenith.

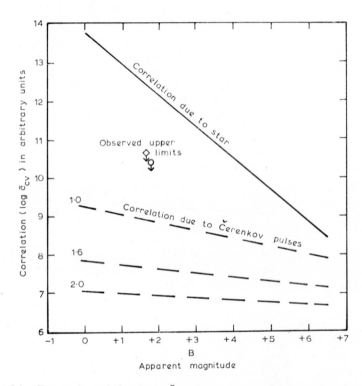

Fig. 5.2. Expected correlation due to Čerenkov pulses for an intensity interferometer, observing a star of apparent magnitude B, is shown by the broken lines for a baseline of 10 m. and three exponents of the integral pulse-height spectrum. The full curve is the correlation due to the star (unresolved). All curves correspond to observations in the zenith with a field of view of 15′ diameter. Experimental upper limits to the correlation from the night sky are shown by the square (11 September 1966) and circle (30 August 1967). From Hanbury Brown, Davis and Allen (1969).

66

An inspection of the results in fig. 5.2 suggests strongly that any correlation due to Čerenkov light would be negligible over the whole working range of the Narrabri instrument ($B < 2.5$). Several tests were made to check this conclusion, and in two of these tests (see §11.10) a blank region of the night sky was observed for many hours to see whether or not there was any unwanted correlation. No such correlation was found and the results of the two tests are shown in fig. 5.2 as experimental upper limits.

Hanbury Brown *et al.* also discuss the effects of Čerenkov pulses on larger and more sensitive instruments than the interferometer at Narrabri. They show that the limits to performance set by Čerenkov light are always likely to be unimportant since they are less restrictive than those set by the shot noise due to background light from the night-sky. For example, from fig. 5.2 we can see that for a field of view 15 minutes of arc in diameter the limiting stellar magnitude, set by uncertainties about Čerenkov radiation, is roughly $+6$, but it is simple to show that the excess shot noise at the correlator output, due to the general background light from the sky, would be about 50 per cent of that due to a star of magnitude $+6$ and would therefore more than double the exposure times. In practice this 'background noise' would reduce the limiting magnitude to about $+4$ or $+5$ where the correlation due to Čerenkov light would be unimportant.

The limits set both by Čerenkov light and by the noise from the sky background can, of course, be improved by reducing the field of view. But the excess noise due to the background light decreases directly as Ω_0, while the unwanted correlation due to Čerenkov light must necessarily decrease as a higher power of Ω_0 due to the factor $f(\Omega_0/\Omega_{cv})$ in equation (5.20). It follows that in any interferometer designed to reach stars fainter than about magnitude $+4$ it will be necessary to reduce the field of view to less than the present field of 15 minutes of arc, and the limits set by shot noise due to the light from the night-sky will always be reached before those set by correlation due to Čerenkov radiation.

5.8 *A Brief Theory of the Effects of Atmospheric Scintillation*

5.8.1 *A mathematical model of scintillation*

A satisfactory theoretical analysis of the expected effects on an intensity interferometer of atmospheric scintillation has not yet been published. Only a brief discussion by Hanbury Brown and Twiss (1958 a) has appeared in print and, in the absence of a more thorough treatment, we shall summarize it here. They consider first the effects of atmospheric scintillation on measurements of an unresolved star. Under these conditions the light at the top of the Earth's atmosphere can be represented by a set of plane waves which in turn can be represented as two independent and orthogonally polarized components. The electric vector in one of these components, the rth, can be represented by a Fourier series of the form

$$E(t) = \sum_{r=1}^{\infty} E_r \cos\left[(2\pi r/T)\,(t-(z-z_0)/c) - \phi_r\right] \qquad (5.23)$$

where the z axis is taken along the line joining the observer and the star and where z_0 is the length of the equivalent vacuum path from the observer to the top of the atmosphere.

At the bottom of the atmosphere an incident plane wave will appear as a continuum of plane waves produced by refraction and diffraction at atmospheric irregularities, so that the electric field at a particular point (x, y, o, t) will have components along all three spatial axes, the lth component being of the general form

$$E_{lr}[1 + \rho_{lr}(\mathbf{x}_1, t) \cos\left((2\pi r t/T) - \phi_r - \psi_{lr}(\mathbf{x}_1, t)\right)] \qquad (5.24)$$

where $\rho_{lr}(\mathbf{x}_1, t)$, $\psi_{lr}(\mathbf{x}_1, t)$ are the fluctuating amplitude and phase variables respectively which determine the scintillation at the point $(\mathbf{x}_1, t)$.

If $(\mathbf{x}_1)$ represents a point on the photocathode of one of the photo-tubes, then the probability of the emission of a photoelectron $p(\mathbf{x}_1, t)$ is proportional to the square of the amplitude of the electric field and they show that

$$p(\mathbf{x}_1, t) \sim \sum_{l=1}^{3} \left\{ \sum_{r=1}^{\infty} \tfrac{1}{2}\alpha_r E_{lr}{}^2 [1 + \rho_{lr}(\mathbf{x}_1, t)]^2 \right.$$

$$+ \sum_{r=s}^{\infty} \sum_{s=1}^{\infty} (\alpha_r \alpha_s)^{1/2} E_{lr} E_{ls} (1 + \rho_{lr}(\mathbf{x}_1, t))\,(1 + \rho_{ls}(\mathbf{x}_1, t))$$

$$\left. \times \cos\left[(2\pi(r-s)t/T) - (\phi_r - \phi_s) - (\psi_{lr}(\mathbf{x}_1, t) - \psi_{ls}(\mathbf{x}_1, t))\right] \right\} \qquad (5.25)$$

where α_r is the quantum efficiency of the photocathode at a frequency r/T.   To find the probability that a photon will be emitted when the light is focused on the photocathode by a parabolic mirror of aperture A, the quantity $p\,(\mathbf{x}_1, t)$ is integrated over the aperture so that

$$P_1(t) = \int_A p(\mathbf{x}_1, t)\, d\mathbf{x}_1. \qquad (5.26)$$

In an intensity interferometer the observed correlation is proportional to the time-average of the joint probability of photoemission at the two photocathodes.   Consideration of equation (5.25) therefore suggests that the correlation observed from the star may be affected by atmospheric scintillation in at least two ways; first, there will be the additional phase-shifts (ψ_{lr}, ψ_{ls}) introduced into the light reaching each photocathode—and these phase-shifts may not be the same at each detector;

68

secondly, the amplitude of the fluctuations in the outputs of the detectors will be modulated by the scintillations. These two effects are considered in turn.

5.8.2. *Phase-scintillation*

Hanbury Brown and Twiss assume that atmospheric scintillation is caused by three-dimensional irregularities of refractive index and that the local deviations are small compared with the average value. Following Bramley (1955) they write the fluctuating phase-shift as

$$\psi(\mathbf{x}_1, t) = \int_0^L (\mu - \mu_{av})\, dz = \int_0^L \delta\mu\, dz \qquad (5.27)$$

where μ, μ_{av} are the instantaneous and average values of the refractive index respectively. Some of the atmospheric irregularities will increase and some will decrease the optical path length so that the effect is that of random walk and the phase of the emergent wave is randomly distributed with a deviation given by

$$\delta\psi(\mathbf{x}_1, t) = (2\pi/\lambda)\,(lL)^{1/2}\delta\mu \qquad (5.28)$$

where $\delta\mu_m$ is the standard deviation of the refractive index; L is the total path length through the atmosphere; l is the size of a typical irregularity; λ is the wavelength of the light.

If the effects of dispersion are neglected, then the phase-scintillation can be represented by a fluctuating time delay τ which is introduced into the light path by the atmosphere. The r.m.s. value of τ is given by

$$(\overline{\tau^2})^{1/2} = (\delta\mu/c)\,(lL)^{1/2}. \qquad (5.29)$$

Hanbury Brown and Twiss take $L = 10$ m, $l = 1$ m and $\delta\mu = 10^{-6}$ which corresponds to an r.m.s. time delay of 0·3 ps. A study of more recent papers on the effects of the atmosphere on the light from lasers (e.g. Hodara, 1966) suggests that these values are reasonable and that even with the most extreme assumptions, the r.m.s. fluctuations of delay are unlikely to exceed 1 ps.

The effect of inserting a time delay in the light reaching one detector has already been discussed in § 4.2.4. Substituting in equation (4.34) it is simple to show that the loss of correlation due to a delay τ in one channel can be written

$$\frac{1}{\Delta f} \int_0^{\Delta f} \cos(2\pi f\tau)\, df \approx 1 - \tfrac{1}{6}(2\pi\Delta f\tau)^2 \qquad (5.30)$$

where Δf is the bandwidth of the electronic system which is assumed to be rectangular for simplicity. If we take the bandwidth as 0–100 MHz then the loss of correlation for a time delay of 1 ps is extremely small, roughly 1 part in 10^7.

69

It follows that the random time delays introduced by atmospheric scintillation have no significant effect on the correlation. Furthermore, it is likely that the value of τ which we have assumed, is extreme, and that we have overestimated the random delay.

Hanbury Brown and Twiss point out that the effects of dispersion have been neglected in their analysis and they argue that they are negligible, as follows. The correlated components in the outputs of the two detectors can be regarded as intermodulation or difference frequencies between frequency components present in the original light wave. If the bandwidth of the electronic circuits is restricted to frequencies less than 100 MHz, then phase-dispersion in the atmosphere can only affect the correlation if it introduces significantly different phase-shifts into components of the original light which differ in frequency by only 100 MHz; furthermore, the amount of this phase-dispersion must be different at the two detectors. Taking the usual formula for the refractive index of air, it is simple to show that the difference in phase introduced by the entire atmosphere into light waves which are 100 MHz apart in frequency is only about 0·5 radians at 400 nm. It is therefore clear that any differential phase-dispersion produced at the two detectors by minor irregularities in the atmosphere will be very much less than 0·5 radian and its effect on the observed correlation will be negligible.

5.8.3 *Amplitude scintillation*

Turning to the effects of amplitude scintillation Hanbury Brown and Twiss assume, for simplicity, that the amplitude scintillations of two light frequencies received at a given point are fully coherent if these frequencies are separated by less than 100 MHz. They argue that this assumption can be justified theoretically and is supported by the observations of Mikesell, Hoag and Hall (1951). Under these conditions it is permissible to write

$$1 + \rho_{lr}(\mathbf{x}_1, t) = 1 + \rho_{ls}(\mathbf{x}_1, t) \tag{5.31}$$

in equation (5.24) and to represent the effects of amplitude scintillation by expressing the probability that a photoelectron is produced by light incident at any point of one mirror as $1 + q_1(t)$ where

$$q_1(t) = \frac{\sum\limits_{r=1}^{\infty} \sum\limits_{l=1}^{3} \int_{A_1} [(1 + \rho_{lr}(\mathbf{x}_1, t)^2 - 1] \, d\mathbf{x}_1 \alpha_r E_{lr}^2}{\sum\limits_{r=1}^{\infty} \sum\limits_{l=1}^{3} \alpha_r E_{lr}^2 A_1} \tag{5.32}$$

and A_1 is the aperture of one of the mirrors.

They argue that the quantity $q_1(t)$ is an irregularly fluctuating quantity with zero mean value and with a power spectrum which in practice seldom extends above about 100 Hz. The r.m.s. value of

$q_1(t)$ depends upon the mirror aperture and the zenith angle of the star. For mirrors of the order of 3 m in diameter and for zenith angles less than 45°, they estimate that $q_1(t)$ will not exceed about 0·03 even under conditions of bad seeing. The effect of this amplitude scintillation will be to increase the correlation by the factor

$$\langle (1 + q_1(t))(1 + q_2(t)) \rangle_{av} = 1 + \langle q_1(t)q_2(t) \rangle_{av} \qquad (5.33)$$

when amplitude scintillation is present. Thus, even in the worst imaginable case, when the amplitude scintillations are completely correlated at the two detectors and when the r.m.s. value of $q_1(t) = 0·03$, the correlation would be increased by less than 1 part in 10^3. In practice the amplitude scintillations at two detectors separated by a few metres will be almost uncorrelated and therefore the effect will be even less. They conclude that the effects of amplitude scintillation on an intensity interferometer are negligible.

It is interesting at this point to refer to the measurements, made at Narrabri with reflectors 6·5 m in diameter, which are described in §11.11. They show that the observed amplitude scintillation was very much smaller than the value $q(t) = 0·03$ assumed above.

5.8.4 *Angular scintillation*

Finally, Hanbury Brown and Twiss consider the observation of a star with a finite angular size. In this case the correlation may be altered if atmospheric scintillation introduces appreciable differential phase-shifts into light waves reaching a point on the Earth from different points on the star. Such differential shifts might be introduced if rays from different points on the star traverse different irregularities. However, this effect cannot be significant if the two points at which the rays traverse the irregularities are so close together that they lie in the same Fresnel zone as viewed from the receiving point. For example, if the angular diameter of the star is 0·01 seconds of arc and the irregularities are at any height up to say 10 km, then the maximum separation between two rays from the extremities of the star cannot be greater than 0·05 cm. At a wavelength of 400 nm the diameter of the first Fresnel zone on a plane at a height of 10 km as seen from a point on the Earth, is of the order of 5 cm. It follows that the effects of these differential phase-shifts are unlikely to be significant.

A second way in which differential phase-shifts could arise is through angular scintillation. Thus it is known from experiment that, for a specific point on the reflector, the instantaneous direction of the incident light can fluctuate by as much as 3 seconds of arc around the mean value with a standard deviation of about 1 second of arc under conditions of rather poor seeing. In the most unfavourable case, where the scattering occurs in a single thin layer, the differential phase-shifts introduced by this effect into light emitted by different points on the

71

star are just those which would arise if the observing point on Earth were to be translated horizontally by a distance

$$\xi = \chi H \sec z \qquad (5.34)$$

where χ is the scattering angle, H is the height of the scintillating layer, z is the zenith angle of the star. Taking as an extreme value $H \sec z = 5 \times 10^4$ m and the standard deviation of χ to be 1 second of arc, then the standard deviation of ξ is about 25 cm.

Experience with large reflectors shows that the correlation between angular scintillations at points more than about 30 cm apart is usually small. Hence Hanbury Brown and Twiss argue that the effect of angular scintillation on an intensity interferometer can be analysed on the assumption that elements of the reflector aperture, of the order of 30 cm in size, are randomly displaced with respect to one another by distances of the order of 25 cm. This movement would effectively change the shape and area of the reflector and randomly displace the optical centre by an amount which is appreciably smaller than the displacement of a single element. For mirrors of several metres diameter this random displacement would probably be less than 25 cm by an order of magnitude; furthermore, the direction of this displacement would be random and its mean value would be zero over a long period. Thus they conclude that, for measurements which involve baselines of the order of several metres, the effect of this random apparent displacement of the optical centres of the reflectors by angular scintillation would be negligible; they also add that it can be shown that the random changes in the effective shape and aperture of the reflectors are also unlikely to produce significant effects.

5.8.5 *Summary of the effects of scintillation*

In summary, Hanbury Brown and Twiss have advanced plausible arguments that the effects of atmospheric scintillation on the correlation observed with a large intensity interferometer are negligible and their analysis is supported by the observations at Narrabri reported in § 11.11. Thus, from a severely practical point of view, we may safely conclude that the effects of scintillation are unlikely to have been significant in the measurements at Narrabri which were carried out over a limited range of zenith angles (less than $60°$) and with an accuracy of only a few per cent. Nevertheless, this freedom from the effects of scintillation is a remarkable and important property of an intensity interferometer and it would be worth while to refine both the analysis and the observations to establish these effects more precisely.

CHAPTER 6

laboratory tests

6.1 *Tests of a Linear Multiplier Intensity Interferometer*

The first laboratory test of an intensity interferometer was carried out in 1955 (Hanbury Brown and Twiss, 1956 a). This test established that when two photoelectric detectors are illuminated with coherent light the fluctuations in their output currents are correlated; the measured correlation was in fair agreement with theory. In view of the controversy which surrounded the publication of these results, it was decided to repeat the experiment with improved apparatus and greater precision. This second experiment was carried out at the Jodrell Bank Experimental Station of the University of Manchester in 1957 (Hanbury Brown and Twiss, 1957 c).

6.1.1 *The optical system*

Fig. 6.1 shows a simplified outline of the optical system used in this second experiment. The light source was formed by a circular pinhole 0·19 mm in diameter on which the image of a mercury arc lamp was focused by a lens. The image of the arc was adjusted so that the pinhole lay in the brightest part close to one of the electrodes. The 435·8 nm line of mercury was isolated by a liquid filter. The beam of light from the pinhole was divided by a semi-transparent mirror to illuminate the cathodes of the photomultipliers P_1, P_2. The area of each cathode exposed to the light was limited by a square aperture of

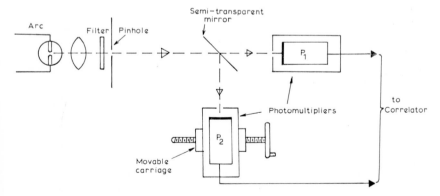

Fig. 6.1. Optical system of correlation experiment.

73

5×5 mm, and the distance from the pinhole to each cathode was adjusted to be 2·240 m.

The photomultipliers were a matched pair (R.C.A. type 6342) with flat end-on cathodes and ten stages of 'focused dynode' multiplication. The cathodes had a maximum response at 400 nm; their spectral responses were almost identical and their quantum efficiencies were 16·9 and 14·6 per cent at 400 nm.

The degree of coherence of the light at the cathodes could be varied by traversing one photomultiplier (P_2) horizontally and normal to the incident light. Thus the cathode apertures, as viewed from the pinhole, could be superimposed or separated by any desired amount up to several times their width.

The fluctuations in the anode currents of the photomultipliers were transmitted to a correlator through coaxial cables of equal length. The d.c. component of these currents was separated by a filter and measured separately.

6.1.2   *The correlator*

The most difficult problem in building an intensity interferometer is to make a satisfactory correlator. At first sight the problem looks reasonably simple; one has only to multiply two wide-band voltages together and observe their product. But the practical difficulties in doing this are severe; the correlation between the two voltages is so small that, unless special precautions are taken, the measurements are seriously affected by very small drifts in the zero of the multiplier output.

In §4.2 we showed that the correlation is the product of two small components of wave noise $\overline{j_c^2}$ which are submerged in much larger uncorrelated components of shot noise $\overline{j_n^2}$. If we now consider observations of a bright star, where the ratio of correlation to r.m.s. noise in the output of the multiplier is typically 1/1 in an observation lasting 100 s, then, for an electrical bandwidth $\Delta f = 100$ MHz, equations (4.18) and (4.19) show that $\overline{j_c^2}/\overline{j_n^2} \approx 10^{-5}$ at the input to the correlator. This corresponds to a correlated component which is 50 dB below noise at the inputs to the multiplier and 100 dB below noise in the output. As a first attempt, in 1955, to solve the problem of measuring this extremely small component, the technique of 'phase-switching' was borrowed from radio-astronomy. A phase-reversing switch was inserted in one input to the multiplier and arranged to reverse the phase of the signal 10 000 times per second. This operation converted the small d.c. component in the output of the multiplier, corresponding to the correlation, into a low-frequency signal which was then amplified in a narrow-band amplifier thereby removing most of the higher-frequency components of noise. The output of this amplifier was synchronously demodulated to give an 'output' which was integrated

74

and recorded. It was found experimentally that this system could not be made sufficiently stable. If two completely uncorrelated wide-band noise voltages were applied to the correlator then, over periods of hours, the integrated output of the synchronous demodulator would tend to drift irregularly away from zero by a significant amount. To combat this zero-drift a second phase-switch was introduced to reverse the phase of the other input to the multiplier every 10 s. The output of the first synchronous demodulator was then passed through a low-pass filter to reduce further the noise and was then 'synchronously demodulated' for the second time by a simple reversing switch operated every 10 s. The output of this final reversing switch was taken directly to the recorder, which in this case was an integrating motor. The object of this second demodulator was to reduce the zero-drift in the output of the first demodulator, and it is a feature of the system that there are no active components in the second demodulator which might introduce drifts. All subsequent correlators have been based on this technique of double phase-switching.

Fig. 6.2 shows a simplified outline of the correlator used during the second laboratory experiment in 1957. The cable from each of the

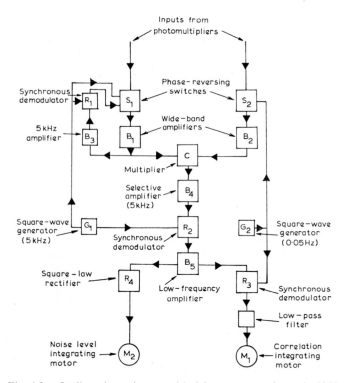

Fig. 6.2. Outline of correlator used in laboratory experiment in 1957.

photomultipliers was terminated in a matched load and the voltage fluctuations across this load were applied to one of the two input channels of the correlator. Both channels consisted of a phase-reversing switch followed by a wide-band amplifier. The phase-reversing switch S_1 in channel 1 was electronic and reversed the phase of the input voltage 10 000 times per second in response to a 5 kHz square wave from the generator G_1. It was essential, because of non-linearities in the multiplier, to reduce amplitude modulation of the signal by this switch to an extremely low level in order to reduce spurious signals at 5 kHz in the output of the multiplier; for this reason the gain of the switch was equalized in both positions by means of an automatic balancing circuit comprising a rectifier (not shown) which detected any amplitude at the output of the amplifier B_1, a selective 5 kHz amplifier B_3 and a synchronous demodulator R_1. The phase-reversing switch S_2 in channel 2 consisted of a relay-operated coaxial switch which reversed the phase of the input every 10 s in response to an 0·05 Hz square wave from the generator G_2. The wide-band amplifiers B_1, B_2 were identical in construction and their gain was substantially constant (± 1 dB) from about 5 to 45 MHz and decreased rapidly outside this band. Their outputs were multiplied together in the multiplier C, which consisted of a balanced arrangement of two pentode valves with their anodes in push–pull. The output of this multiplier was amplified by a high-gain selective amplifier B_4 tuned to 5 kHz with a bandwidth of 70 Hz. The output of B_4 was applied to the synchronous demodulator R_2 which consisted of a ring of diodes synchronized by the 5 kHz wave generated by G_1. The demodulator R_2 was followed by an 0·05 Hz amplifier B_5 which was relatively broad-band and passed all frequencies from 0·01 to 0·25 Hz. The final synchronous demodulator R_3 consisted of a relay which, in response to the 0·05 Hz square wave from G_2, periodically reversed the connections between the output of the amplifier B_5 and the integrating motor M_1. A low-pass filter, containing only passive elements, was inserted between the output R_3 and the integrating motor to restrict the bandwidth of the noise to the range 0 to 0·01 Hz. The integrating motor was a miniature motor coupled to a revolution counter; it was capable of rotation in either direction and tests showed that the relation between speed and input voltage was linear to better than 1 per cent. An additional integrating motor M_2 was provided to monitor the r.m.s. level of output voltage from the amplifier B_5.

If now the photomultipliers in this arrangement are illuminated with uncorrelated beams of light ($\gamma_{12} = 0$), then the output of the multiplier is random noise with a spectral density which has a maximum around zero frequency and which decreases to zero at about 40 MHz. The corresponding output from the amplifier B_4 is random noise centred about 5 kHz with a bandwidth of 70 Hz. After passing through the synchronous demodulator R_2 the spectrum extends from 0 to 35 Hz, and after passing through amplifier B_5 and the second synchronous demodulator R_3, it is reduced to a band extending from about 0 to 0·25 Hz. The low-pass filter following R_3 finally restricts the bandwidth to the range 0 to 0·01 Hz. Under the influence of this noise the motor M_1 spins in either direction at random and the reading of the revolution counter remains close to zero. However, if the light is coherent at the photomultipliers ($\gamma_{12} > 0$), a 5 kHz component appears at the output of the multiplier; this component is coherent with the 5 kHz switching wave and reverses in phase every 10 s in synchronism with the 0·05 Hz switching wave. After amplification by B_4 the 5 kHz component produces a 0·05 Hz square wave in the output of the synchronous demodulator R_2; this in turn is amplified by B_5 and is synchronously rectified by R_3 to produce a d.c. component in the voltage applied to the integrating motor. Thus, when the light is coherent at the two photomultipliers, the integrating motor revolves more in one direction than the other and the reading on the revolution counter increases with time as a measure of the correlation.

Any unwanted coupling between the two input channels, or unwanted external signals, introduces spurious correlation if it occurs before the phase-reversing switches. It was therefore necessary to screen the photomultipliers very thoroughly against coupling or electrical interference.

Extensive tests of this correlator, using independent light sources to illuminate the phototubes, showed that over several hours the zero drift in the output was less than the theoretical uncertainty due to noise alone. Although short-term deviations, greater than would be expected, were in fact observed, it was found that their effect was negligible in observations lasting more than 30 min.

6.1.3 *Experimental procedure and results*

The two photocathodes, as viewed from the light source, were superimposed by adjusting the position of the movable photomultiplier P_2 (fig. 6.1). Readings were then taken every 5 min, for a total period of 4 hours, of the revolution counters on the integrating motors M_1 and M_2 and also of the anode currents of the photomultipliers. The centres of the two photocathodes, as seen from the light source, were then separated by 1·25, 2·50, 3·75 and 10·0 mm. In each of these positions readings were taken at 5 min intervals for about 30 min; the readings were then repeated with the cathodes separated by the same distances but in the opposite direction.

Throughout the experiment the gain of the amplifier B_4 (fig. 6.2) was controlled to keep the output noise from the correlator approximately constant. The gains of the two photomultipliers were measured before and after every run. Table 6.1 shows the experimental results. The correlation and noise measured in each 5 min interval were weighted appropriately by the anode currents and added to give an r.m.s. signal/noise ratio for each cathode separation. These ratios are independent of the inevitable small changes in the light from the arc lamp and of changes in the gain of the correlator. They are shown in column 2 of Table 6.1.

Cathode separation d/mm	Observed correlation Signal/noise (r.m.s.)	Theoretical correlation Signal/noise (r.m.s.)
0	+17·55	+17·10
1·25	+8·25	+8·51
2·50	+5·75	+6·33
3·75	+3·59	+4·19
5·00	+2·97	+2·22
10·00	+0·90	+0·13

Table 6.1. The experimental and theoretical correlation between the fluctuations in the outputs of two photoelectric detectors illuminated with partially coherent light. From Hanbury Brown and Twiss (1957 c).

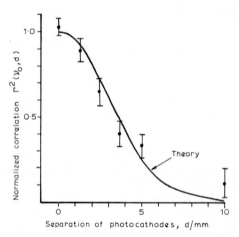

Fig. 6.3. The experimental and theoretical values of the normalized correlation factor $\Gamma^2(\nu_0, d)$ for different separations between the photocathodes. The experimental results are shown as points with their associated probable errors. From Hanbury Brown and Twiss (1957 c).

6.1.4 *Comparison between theory and experiment*

The theoretical signal/noise ratios for each cathode spacing were calculated from equation (5.17) and are also shown in Table 6.1. The parameters A_1, A_2, σ, β_0, $g(\nu_0)$, $\alpha(\nu_0)$, $n(\nu_0)$, b_v, η, $(1+a)$, $(1+\delta)$, μ were measured in the laboratory and ϵ, $\Delta(\nu_0)$ and $\Gamma^2(\nu_0, d)$ were calculated. The results are also shown in fig. 6.3 in a form which shows clearly how the correlation decreased with cathode spacing. In this figure the observed signal/noise ratios have been normalized by the theoretical values for zero cathode spacing. Effectively this procedure yields experimental values for the normalized correlation factor $\Gamma^2(\nu_0, d)$ and the results have been plotted in fig. 6.3 for comparison with the theoretical values of this function shown as a solid curve.

The data show that the experimental results are in agreement with theory within the uncertainties set by statistical fluctuations. In principle the comparison could have been made closer by increasing the time of observation. However, in practice, it is unlikely that this particular experiment could have been made much more precise without considerable elaboration. It would soon have reached limits set by uncertainties in some of the parameters. In this connection we note in retrospect that the time delays in the two phototubes were not measured and equalized, as in later interferometers; experience suggests that there might have been an unsuspected loss of correlation of as much as 5 per cent. Furthermore, it must be remembered that the theoretical signal/noise ratios are themselves based on a number of simplifying assumptions; for example, it is assumed that the quantum efficiency is constant over the photocathodes, and these approximations might well have prevented any substantial increase in the precision of the comparison between theory and experiment.

6.2 Tests of a Coincidence-counting Intensity Interferometer

6.2.1 Introduction

The first attempt to test whether there is any correlation between the times of arrival of photons in coherent beams of light appears to be that made in Hungary by Ádam, Jánossy and Varga (1955). They split a light beam into two parts with a semi-transparent mirror and used two photomultipliers to look for time-coincidences between photons in the two beams. They found no evidence for excess coincidences and concluded that "the number of photons taking part in coincidences is certainly not larger than 0·6 per cent". They argued that the photons of a light beam which is being split into two components will be contained in either the one or the other component and that "if these two components fall on the cathodes of two photomultipliers, the individual photons contained in either light component can be recorded and, according to quantum theory, the pulses recorded by the two multipliers will be independent of each other; thus no systematic coincidences are to be expected".

Following the publication by Hanbury Brown and Twiss (1956 a) of the results described earlier in this chapter a second attempt was made to detect the correlation between photons using a coincidence counter. This second attempt was made in Canada by Brannen and Ferguson (1956). They repeated the experiment of Ádám et al., but with improved equipment, and concluded again that "there is no correlation between photons in coherent light rays''. To be more precise, their experiment showed that the probability that more than 0·03 per cent of the photons were in true time coincidence was less than one part in 10^7. They claimed that, "if such a correlation did exist, it would call for a major revision of some fundamental concepts in quantum mechanics".

It is worth while to recall these two experiments because they illustrate how misleading the mental picture of photons can be. In both cases the objection to the correlation between photons stemmed from the idea that one can only expect time-correlation from bits of the same photon and, since quantum theory does not allow us to split photons, no correlation is to be expected.

Although the theoretical interpretation of these experiments was completely wrong there is no reason to reject the experimental results. It can be shown theoretically (Hanbury Brown and Twiss, 1956 c) that in the experiment of Ádám et al. the fraction of photons in true coincidence should be less than 10^{-9}, which is consistent with their experimental upper limit of 6×10^{-3}. In the second experiment, of Brannen et al., the theoretical expectation is about 10^{-8}, which is again consistent with their experimental upper limit of 3×10^{-4}. As we have already shown in §4.3.3, a successful experiment of this type can only be performed with a bright source of light with a very narrow spectral distribution. In neither of these experiments was this condition fulfilled and consequently they could not be expected to yield significant results.

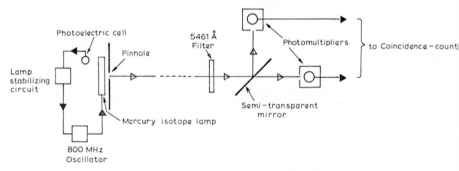

Fig. 6.4. Arrangement of the optical system in the experiment of Twiss and Little.

6.2.2 *The experiment of Twiss and Little*

The first demonstration of the time-coincidence between individual photons was reported by Twiss, Little and Hanbury Brown (1957). This experiment was carried out in Australia and has been described in detail by Twiss and Little (1959). The arrangement of the optical system is shown in fig. 6.4. The light source consisted of an electrodeless mercury isotope lamp excited at 800 MHz. The visible area of the light source was limited by a circular pinhole 0·36 mm in diameter in a brass tube which fitted tightly over the discharge tube. The light flux from this lamp was 0·0013 W cm^{-2} sr^{-1} and the coherence time was measured to be $\tau_0 = 0.73$ ns. The 546·1 nm line of mercury was isolated by a filter and the beam was split by a semi-transparent mirror to illuminate two photomultipliers. The areas of the photocathodes exposed to the light were limited by square apertures 2×2 mm at a distance of 1·25 m from the source. Both phototubes were mounted on movable slides so that, as seen by the source, they could be optically superimposed or separated by a distance of 5 mm transverse to the line of sight; at this latter separation the pinhole source was completely resolved, so that the incident light beams were uncorrelated. The outputs of the phototubes were taken to a coincidence counter with a resolving time of 3·5 ns. In order to establish the random counting rate a delay of 15 ns could be inserted in one channel.

The measurements were carried out as follows: the photocathodes were optically superimposed and the number of coincidences in 2 min n_{1r} was recorded; one of the photocathodes was then moved to the displaced or uncorrelated position and the number of coincidences in 2 min n_{2r} was again recorded. The whole procedure was repeated ten times in a single run which took about 50 min to complete.

If the average light intensity reaching the movable photocathode had been equal in the coincident and displaced positions, then $(N_1 - N_2)/N_2$ would have given a measure of the ratio of the excess or 'correlated' coincidences to the random coincidences, where

$$N_1 = \sum_{r=1}^{10} n_{1r}, \quad N_2 = \sum_{r=1}^{10} n_{2r}. \tag{6.1}$$

However, in this experiment there was a difference of the order of 0·5 per cent between these two intensities and, to eliminate the effect of this, a comparison run with the delay cable in one channel of the coincidence counter was made after each observation. The delay in this cable being about four times the resolution of the coincidence counter, there was no chance of a count being registered by the simultaneous arrival of photons at the two photocathodes.

The procedure in the comparison run was identical with that in the observation run. From this second set of results a correction factor ϵ was computed, where

$$\epsilon = \sum_{r=1}^{10} n'_{2r} \bigg/ \sum_{r=1}^{10} n'_{1r} = N_2'/N_1' \tag{6.2}$$

and ϵ represents the ratio of the light fluxes incident upon the movable photocathode in the displaced and coincident positions. The number

81

of excess or 'correlated' coincidences n_c, corrected for differences in light intensity, is then given by

$$n_c = \rho_c N_2 \tag{6.3}$$

where ρ_c is the ratio of correlated to random coincidences and is given by

$$\rho_c = (N_1 \epsilon - N_2)/N_2. \tag{6.4}$$

A total of six runs, amounting to 480 min of observation, was made using the procedure described above and the results are shown in Table 6.2. Combining these results, making a total number of coincidences of about 7×10^5, the experimental ratio of correlated to random counts is

$$\rho_c(\text{experimental}) = 0 \cdot 0193 \pm 0 \cdot 0016 \text{ (p.e.).} \tag{6.5}$$

As a final check a seventh run was made with a tungsten filament lamp in place of the isotope lamp; as expected, no significant correlation was observed as shown by the bottom line in the Table.

The theoretical value of ρ_c is given by equations (4.51) and (4.53) and it can be shown that

$$\rho_c = (\tau_0/4\tau_c)\Delta(\nu_0)g \tag{6.6}$$

where τ_c is the resolving time of the coincidence counter; $c\tau_0$ is the coherence length of the light; g takes account of a number of small losses of correlation in the equipment due, for example, to polarization effects in the semi-transparent mirror; $\Delta(\nu_0)$ is the partial coherence factor (see §5.3) which allows for the finite size of the photocathode. Taking the measured values $\tau_0 = 0 \cdot 73$ ms, $\tau_c = 3 \cdot 5$ ns, $g = 0 \cdot 86$ and the

Run	Random coincidences	Correlated coincidences	Correlated/Random $\rho_c \times 100$
1	135 446	2986	2·20
2	128 975	2190	1·70
3	136 250	2369	1·73
4	99 640	2185	2·19
5	96 326	1864	1·94
6	93 848	1761	1·88
Dummy run with white light	81 576	185	0·22

Table 6.2. Ratio of correlated to random coincidences of photons observed by Twiss and Little (1959).

calculated value $\Delta(\nu_0) = 0.475$, Twiss and Little found the theoretical value of ρ_c to be

$$\rho_c(\text{theoretical}) = 0.0207. \tag{6.7}$$

The uncertainty in this figure, due to uncertainties in the parameters, was estimated to be about ± 0.002.

A comparison of the theoretical and experimental values of ρ_c shows that they are in satisfactory agreement, the discrepancy being less than the uncertainty in either value. Thus the experiment confirms that the arrival times of photons at different points *are* correlated when these points are illuminated by mutually coherent beams of light.

Twiss and Little carried out two further experiments with their coincidence-counting interferometer. In one experiment they measured the variation of photon-correlation as a function of the displacement of the photocathodes. The results were in good agreement with theory. In another experiment they measured the coincidence rate with the two cathodes superimposed, but with the two light beams alternately parallel and orthogonally polarized. This latter experiment confirmed that photons are correlated when the beams are parallel polarized but uncorrelated when they are orthogonally polarized.

6.2.3 *Other experiments*

Following the work of Twiss and Little, the correlation between photons was confirmed, using coincidence counters, by Rebka and Pound (1957), by Brannen, Ferguson and Wehlau (1958), by Martienssen and Spiller (1964) and by Farkas, Jánossy, Náray and Varga (1965). All these experiments have confirmed that the theory of the correlation between photons, as presented in § 4.3 is correct. Although there seems to be little practical application for a coincidence-counting interferometer, it is a striking and easily appreciated demonstration of the wave-particle duality of light.

CHAPTER 7

two early intensity interferometers

7.1 *A Radio Intensity Interferometer*

We have already noted in chapter 1 that the first intensity interfero-meter was a radio interferometer proposed in 1949 specifically to measure the angular sizes of the two most intense radio sources in the sky, Cygnus A and Cassiopeia A. A pilot model was built in 1950 and was tested by measuring the angular diameter of the Sun at 125 MHz. As this first test proved satisfactory a full-scale instrument was built in 1951 and was used successfully to measure the angular sizes of the two radio sources. The theory of this instrument has been published by Hanbury Brown and Twiss (1954); a description of the instrument and the results has been given by Hanbury Brown,

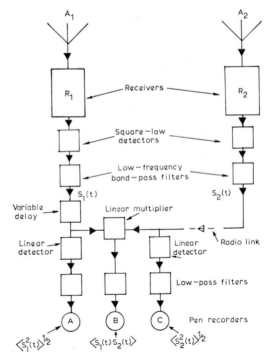

Fig. 7.1. Outline of a radio intensity interferometer.

Jennison and Das Gupta (1952) and, in greater detail by Jennison and Das Gupta (1956).

A simplified block diagram of the equipment is shown here in fig. 7.1. Two spaced aerial systems A_1, A_2, each with an area of 500 m², were connected to two *completely independent* superheterodyne receivers R_1, R_2. These receivers were both tuned to 125 MHz with a bandwidth of 200 kHz. Their intermediate-frequency outputs were rectified in square-law detectors and fed to two identical filters with bandpasses extending from 1 to 2·5 kHz. The low-frequency outputs of these filters were then brought together using a radio link, and their product or correlation measured in a linear multiplier or correlator. To compensate for the time taken for one signal to travel along the baseline and for the difference in the time of arrival of the signals at the two aerials when the direction of the source was not normal to the baseline, an adjustable delay was inserted in one channel. The outputs of the two filters $S_1(t)$, $S_2(t)$ were rectified in linear detectors, giving $\langle S_1^2(t) \rangle^{1/2}$, $\langle S_2^2(t) \rangle^{1/2}$, and recorded by pen recorders. The output of the multiplier $\langle S_1(t)S_2(t) \rangle$ was also recorded. From these three records the normalized correlation $c_N(d)$ was found from

$$c_N(d) = \frac{\langle S_1(t)S_2(t) \rangle}{[\langle S_1^2(t) \rangle^{1/2} - P_{N1}][\langle S_2^2(t) \rangle^{1/2} - P_{N2}]} \tag{7.1}$$

where d is the baseline between the aerials; P_{N1}, P_{N2} are the average values of the filter outputs when the source is not in the aerial beam, or in other words, they are the sum of aerial and receiver noise. Fig. 7.2 illustrates the passage of a source through the beam of the aerials.

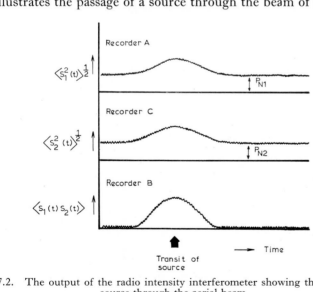

Fig. 7.2.   The output of the radio intensity interferometer showing the transit of a source through the aerial beam.

85

In their analysis of this radio interferometer Hanbury Brown and Twiss (1954) showed that the normalized correlation $c_N(d)$ is simply equal to the square of the fringe visibility in a Michelson interferometer. This result is, of course, to be expected because we have already shown in §3.5 that, for a semi-classical model of photoelectric emission, the correlation is proportional to the square of the degree of coherence $|\gamma_d(0)|^2$. Since this model is precisely equivalent to the radio case, the normalized correlation $c_N(d)$ must also be equal to $|\gamma_d(0)|^2$ and hence, from equation (4.7), to the square of the fringe visibility in Michelson's interferometer. It follows from the discussion in §4.1 that, by measuring $c_N(d)$ as a function of d, the angular size of a radio source can be found.

There is however, one significant difference between the analysis of the radio and optical interferometers which should be noted. The ratio of wave noise to shot noise in the outputs of the detectors is very different in the two cases. This is to be expected because the radio photon carries much less energy than the optical photon. Thus from equation (4.20) the ratio of wave noise $\overline{j_c^2}$ to shot noise $\overline{j_n^2}$ is

$$\overline{j_c^2}/\overline{j_n^2} = \alpha n_0 A \qquad (7.2)$$

where α is the quantum efficiency; A is the area of the detector and n_0 is the number of photons received per unit time, area and bandwidth. For the same energy flux the number of photons n_0 at 100 MHz will be roughly 10^6 greater than at optical wavelengths and, substituting practical values in equation (7.2), we find that in the radio interferometer $\overline{j_c^2} \gg \overline{j_n^2}$ and therefore wave noise predominates; in the optical interferometer $\overline{j_c^2} \ll \overline{j_n^2}$ and shot noise predominates. This difference affects only the calculation of the signal/noise ratio and means that equation (4.30), which we derived for the optical case, cannot be applied to the radio case. The signal/noise ratio of the radio interferometer is more difficult to calculate due to the correlation between the wave noise in both channels; this point is discussed by Hanbury Brown and Twiss.

Successful measurements of both Cygnus A and Cassiopeia A were made with this first radio interferometer and were reported in detail by Jennison and Das Gupta (1956). Cassiopeia A proved to be a radially symmetrical source with an angular diameter of several minutes of arc, while Cygnus A proved to be a double source. The variation of the normalized correlation coefficient with baseline, as observed along the major axis of Cygnus A is shown in fig. 7.3. This result was interpreted as being due to a double source consisting of two equal components with a diameter of about 45 seconds of arc and a separation of $1' 25''$. The maximum baseline used in the experiment was about 12 km.

An interesting feature of this work was that, on occasions, the intensity of the source fluctuated rapidly due to irregularities in the

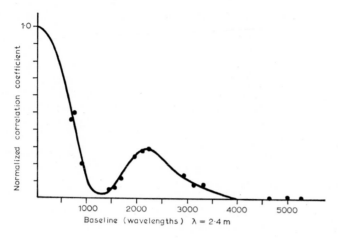

Fig. 7.3. The variation of the normalized correlation coefficient for the source Cygnus A observed at 2·4 m (125 MHz) along the major axis. From Jennison and Das Gupta (1966).

ionosphere. This led us to investigate, both theoretically and experimentally, whether or not the observations of an intensity interferometer are affected by scintillation. A brief account of the theoretical analysis is given by Hanbury Brown and Twiss (1954). They consider first the effects of differential delays or phase-changes introduced by the ionosphere into the signals reaching the two aerials. They estimate that the total delay in the ionosphere is about 1 μs and therefore, any differential delays are likely to be less than this value. Since the bandwidth of the fluctuations reaching the multiplier was limited to 1–2 kHz, it follows from equation (4.34) that such delays would have a negligible effect on the measured correlation. There are also the large fluctuations in intensity due to scintillation and they argue that these will not affect the *normalized correlation*, provided that their period is long compared with the response time of the output integrating circuits. In practice the response time of the integrating circuits was very much less than the period of the fluctuations and so these fluctuations of intensity should not affect the correlation. To summarize, their analysis concludes that the normalized correlation should be independent of ionospheric scintillation. This conclusion is supported by the observations of Jennison and Das Gupta who state that the accuracy of their measurements was not seriously affected by the presence of severe scintillations.

Another interesting point raised by Hanbury Brown and Twiss in discussing this work is the effect of the rotation of polarization, Faraday rotation, in the ionosphere. A plane wave of angular frequency ω travelling parallel to the Earth's magnetic field through a uniform

87

ionosphere of thickness H will be rotated in polarization, relative to a similar wave in free space by

$$Hω_c{}^2ω_H/cω^2 \text{ radians} \qquad (7.3)$$

where $ω_c$ is the critical frequency, $ω_H$ is the gyro frequency and it is assumed that $ω \gg ω_H$ and $ω^2 \gg ω_c{}^2$. They estimated that the rotation of polarization at 100 MHz is of the order of 15 radians. If, therefore, there are significant differences between the magnetic fields or the ionosphere at the two detectors there will be a differential rotation of polarization of the waves received from the source. If this difference is $φ$ radians, the normalized correlation will be reduced by $\cos^2 φ$ in an interferometer using two parallel linearly polarized aerial systems. It is to be expected that over short baselines of a few kilometres this effect would be negligible, but it would be significant over very long baselines. It can be reduced either by the use of higher frequencies or by making the aerial systems sensitive to both planes of polarization.

7.2 *An Optical Stellar Intensity Interferometer*

7.2.1 *The equipment*

As we have just seen, the first test of an intensity interferometer was made at radio wavelengths and was completely successful. Following this, a laboratory test was made in 1956 to verify that the same principle could be applied at light waves and that the correlation could be predicted satisfactorily by a semi-classical model of photoelectric detection. The next step was to make a practical working interferometer, not only to demonstrate the technique, but also to verify that it would work in the presence of atmospheric scintillation. A small pilot model was built which for reasons of economy was only large enough to measure the brightest star in the sky, Sirius. The work of building and testing this instrument was carried out at Jodrell Bank in 1955 and 1956 (Hanbury Brown and Twiss, 1956 b, 1958 b).

A simplified diagram of the equipment is shown in fig. 7.4. The optical system consisted of the two mirrors A_1, A_2 which focused light on to the cathodes of the photomultipliers P_1, P_2. The mirrors were

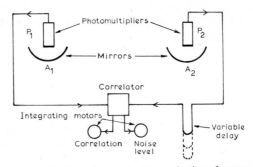

Fig. 7.4. Outline of an optical intensity interferometer.

88

the reflectors of standard Army searchlights and were back-silvered paraboloids of borosilicate glass 156 cm in diameter and 65 cm in focal length; optically speaking, they were crude, and tests showed them to be capable of focusing light from a star into a patch about 8 mm in diameter. The mirrors were supported in their standard searchlight barrels on altazimuth mounts and are shown in fig. 7.5. For these experiments the front glass of each searchlight was removed and the barrel itself was extended with aluminium sheet to form a tube about 2 m long in front of each mirror to exclude extraneous light and to minimize the formation of dew on the mirrors. Even so, experience showed that these 'dew-caps' were not enough, and on cold nights there was often condensation and occasionally ice on the mirrors. This is a trouble not experienced by most astronomical telescopes which are usually protected by a dome and are mounted in some place with lower humidity than the environs of Manchester. However, in the present case, where there was no question of forming an image of the star, the trouble was cured by mounting a 1 kW electric heater in each search-light barrel directly under the mirror. Although somewhat unconventional in terms of normal astronomical practice, this scheme worked well.

Fig. 7.5. The first stellar intensity interferometer at Jodrell Bank (University of Manchester) in 1956.

The azimuth and elevation of the two searchlights were controlled manually from a standard Army control pillar carrying an optical sight; but, to reduce the tedious work of guiding the system entirely by hand, an electric motor was used to rotate the pillar at approximately the correct rate in azimuth.

The arc lamps were removed from the searchlights and photo-multipliers were mounted in their place; the boxes containing them were mounted on sliding carriages so that they could be accurately centred and focused and were heavily screened to prevent the reception of radio interference. The photocathodes were exposed through circular apertures 2·5 cm in diameter and, when observations were not in progress, they were covered by detachable caps carrying small lamps to simulate the light from the star. The photomultipliers had flat semi-transparent cathodes and ten stages of multiplication (R.C.A. type 6342); their spectral response was a maximum at about 400 nm and, at that wavelength, their quantum efficiencies were roughly 15 per cent. The anodes of the photomultipliers were connected by coaxial cables to an electronic correlator in an adjacent laboratory.

In the laboratory the two coaxial cables were brought to a panel where the relative delay of the signals could be adjusted to an accuracy of about 2 ns by lengths of cable which could be plugged into either channel; in this way it was possible to compensate for the varying difference in the times of arrival of the light at the two mirrors as the direction of the star changed throughout the night.

After the delays had been equalized the signals were connected to the inputs of a correlator. This correlator has been described in §6.1. As before, the correlation was recorded by an integrating motor and the r.m.s. value of the noise at the output was recorded by a second integrating motor. The readings of both these motors depended in the same way on the gain of the equipment and the effects of changes in gain were eliminated by expressing all the measurements as signal/noise ratios, that is to say, as the ratio of the time-averaged value of the correlation to the r.m.s. noise or uncertainty in the final value.

7.2.2 *Experimental procedure and results*

Observations of Sirius were made at Jodrell Bank on all possible occasions between November 1955 and March 1956. Only about half of the clear nights could be used, due to moonlight, and the observing time was further limited by the need to avoid excessive atmospheric extinction at low angles. In the case of Sirius, which reaches a maximum of 20° elevation at Jodrell Bank, this limited observations to within 2 hours of transit. Even so, that part of England is not noted for clear skies and the extinction at 20° elevation was frequently excessive; no observations were made when the measured extinction exceeded the value for a clear sky by more than 0·75 magnitudes. The total observing time over a period of 5 months was 18 hours. In addition, about 6 hours of observing time were lost due to failure of the equipment.

The first measurements were made with the two searchlights at the minimum possible separation of 2·56 m. To achieve such a short baseline the searchlights were mounted on a NS baseline so that, at transit, one was looking over the top of the other. Subsequently, observations were made with EW baselines of 5·35, 6·98, and 8·93 m; because the projected baseline varies with the direction of the star, these values are averages taken over the whole observing period.

At each position the searchlights were guided on Sirius and every 5 min readings were taken of the two integrating motors and of the anode currents of both photomultipliers. Measurements were also made of the light contributed by the night-sky. Before and after every run the correlator was checked for sensitivity and drift.

The signal/noise ratios observed at each baseline are shown in Table 7.1.

Baseline /m	Observed signal-to-noise ratio	Theoretical signal-to-noise ratio
2·56	$+8·50 \pm 0·67$	$+9·58$
5·35	$+3·59 \pm 0·67$	$+3·60$
6·98	$+2·65 \pm 0·67$	$+2·69$
8·93	$+0·83 \pm 0·67$	$+1·59$

The baseline is projected normal to the direction of the star and averaged over the run. The signal-to-noise ratios are the ratio of the correlation to the r.m.s. uncertainty in the correlator output.

Table 7.1. Theoretical and experimental correlation from Sirius. From Hanbury Brown and Twiss (1958 b).

7.2.3 *Comparison between theory and observation*

In order to compare these results with theory, Hanbury Brown and Twiss (1958 b) first made an independent estimate of the angular diameter of Sirius ($6·9 \times 10^{-3}$ seconds of arc) and then calculated the signal/noise ratio expected at each baseline. They made these calculations for the actual values of light flux, exposure time and zenith angle appropriate to each 5 min interval. The analysis was greatly complicated by the fact that the optical bandwidth was wide and extended over 200 nm because it had been impracticable to use interference filters. It was therefore necessary to take into account changes in the spectral transmission of the atmosphere with zenith angle and the consequent changes in the spectral density σ (equation (5.6)) and correlation factor $\Gamma^2(d)$ (equation (5.8)). The signal/noise ratios for each interval were evaluated from equation (5.16) using the appropriate values of light flux, spectral density, correlation factor and projected baseline. The variation of the partial coherence factor Δ with spectral response was neglected, as it was too small to be significant. The final

91

signal/noise ratios were found by summing the values for each 5 min interval, weighted by the square of their individual signal/noise ratios. The formulae are too cumbersome to reproduce and are given by Hanbury Brown and Twiss (1958 b). The results of these calculations are shown in Table 7.1.

A comparison of the figures in Table 7.1 shows that for the shortest baseline (2·56 m) the observed signal/noise ratio was $8·50 \pm 0·67$ and the theoretical value was $+9·58$. In the context of this experiment the difference between these values cannot be regarded as significant; the difference is roughly 10 per cent and the uncertainty in the theoretical value cannot be much less. Furthermore, no precautions were taken to equalize the time delays in the two phototubes and subsequent experience suggests that this alone might have been responsible for the loss of several per cent in the observed signal/noise ratio.

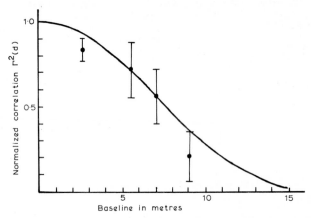

Fig. 7.6. Correlation versus baseline for Sirius. The full line shows the theoretical variation of $\Gamma^2(d)$ the normalized correlation for an angular diameter of $6·9 \times 10^{-3}$ seconds of arc. The points show the experimental results together with their probable errors. From Hanbury Brown and Twiss (1958 b).

The theoretical values for the other three baselines are also shown in Table 7.1 and are in reasonable agreement with the observations. However, a more satisfactory way of showing the variation of correlation with baseline is shown in fig. 7.6. For each baseline the measured values of correlation were normalized using the relations in equation (5.14) to standard conditions of atmospheric transmission and observing time. They were then expressed as a fraction of the theoretical correlation expected for these standard conditions with zero baseline. This procedure converts the observed values of absolute correlation into dimensionless fractions of the correlation to be expected at zero baseline and, since for this experiment the partial coherence factor $\Delta \approx 1$, they can be compared directly with the theoretical values of the correlation factor $\Gamma^2(d)$. The experimental results are plotted

92

in fig. 7.6 together with their r.m.s. errors; the full line shows the theoretical variation of $\Gamma^2(d)$ for Sirius, taking the angular diameter to be $6 \cdot 9 \times 10^{-3}$ seconds of arc. It can be seen that, within the large uncertainties of the experiment, the correlation observed from Sirius did actually decrease with baseline as one would expect theoretically.

An experimental value for the angular diameter of Sirius of $(7 \cdot 1 \pm 0 \cdot 55) \times 10^{-3}$ seconds of arc was also derived by finding the best fit to the results. This value has been corrected theoretically for limb-darkening (see § 10.3.2) and can be compared with the latest value of $(5 \cdot 89 \pm 0 \cdot 16) \times 10^{-3}$ seconds of arc measured at Narrabri. In view of the practical difficulties of this early experiment, and the fact that it was done with almost 'white light', it is not surprising that these two results do not agree more closely.

Finally it must be noted that throughout these experiments Sirius was never more than $20°$ above the horizon and was seen to be scintillating strongly.

7.2.4 *Discussion and conclusion*

The first radio intensity interferometer, described in the first part of this chapter, demonstrated the principle of an intensity interferometer and also made a significant contribution to the study of the radio sources. It also drew attention to one of the principal advantages of the technique, the ability to work through a strongly scintillating medium.

The first optical intensity interferometer, described in the second part of this chapter, confirmed that the technique could contribute to optical astronomy, and drew attention to both of its principal advantages—the relative ease with which one can achieve extremely high resolving power and the ability to work through strong atmospheric scintillation.

The optical experiment had many obvious weaknesses; for example, it did not allow a comparison between the correlation observed with and without scintillation. Nevertheless it was a convincing demonstration and allowed us to proceed with confidence to raise the money for the construction of a full-scale instrument, the stellar intensity interferometer at Narrabri Observatory.

CHAPTER 8

the Narrabri stellar interferometer

8.1   *General Layout*

The general layout of the interferometer at Narrabri is shown in fig. 8.1   The photoelectric detectors were each mounted at the focus of two very large reflectors carried on trucks running on a circular railway track with a gauge of 5·5 m and a diameter of 188 m.   These mobile trucks were connected to the control building by cables suspended from steel catenary wires which were attached at one end to a bearing at the top of a tower in the centre of the circle, and at the other to a small tender towed by each truck.   When not in use the reflectors were housed in a garage built over the southern sector.   A valuable but expensive feature of this garage was a slot running almost the full length of one wall enabling the trucks to be parked inside without detaching the cables and hence without disturbing the electrical connections.

The control building which housed the control desk (fig. 8.7), the computer, a large air-conditioning plant and various motor generators, switchboards, etc., was a two-storey brick building of solid construction with a good heat-reflecting roof so that it was possible even in the summer at Narrabri to hold the inside temperature to $72 \pm 2°$ F.   It was close enough to the central tower to allow the catenary cables to pass over its roof.

8.2   *The Reflectors*

The reflectors (figs. 8.2, 8.5) were regular 12-sided polygons roughly 6·5 m, in diameter, each having a usable reflecting area of 30 m². They were mounted on turntables carried by the trucks and were capable of three independent motions; they could travel around the circular railway track, tilt in elevation about a horizontal axis and rotate about a vertical axis on their turntables.   These three motions were driven by electro-hydraulic motors remotely controlled by servo amplifiers.   To ensure smooth rolling motion on the track the wheels of the trucks were shaped as sections of a cone with its apex at the centre of the track; on the inside rail the wheels were simple rollers but those on the outside track had flanges.   A great deal of trouble was taken to align these wheels very precisely; even so, the inevitable 'scrubbing' action of the motion on a circular track soon ruined the

94

(a)

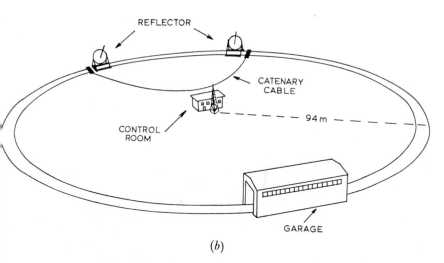

(b)

Fig. 8.1. (a) An aerial view of the stellar interferometer at Narrabri Observatory.
(b) The general layout of the interferometer.

95

surface of the original cast steel and we were compelled to make new wheels of extremely hard steel.

The small tenders towed by the trucks took the radial pull of the catenary and, to do this, they were equipped with side-thrust wheels with vertical axles running on the side of the rails. The tenders also carried auxiliary electronic equipment such as the servo amplifiers, and a manual control console which was used when driving the reflectors in and out of the garage. Since the reflectors were remotely controlled and could not be seen by the operator when working at night, we included an elaborate array of safety devices to prevent accidental damage; for example, long probes extended in front of the trucks to warn against collision and a system of interlocks depending on the separation of the two reflectors and their distance from the garage prevented the long focal probes of the reflectors from fouling each other or the garage.

The framework supporting the reflecting surface was made of light alloy and its bowl was paraboloidal so that all the light from the mirrors reached the focus at the same time. Any differences between the path lengths from the points on the reflecting surface to the focus were small, probably less than 2 or 3 mm. The surface of each reflector was a mosaic of 252 hexagonal glass mirrors, each approximately 38 cm between opposite sides and 2 cm thick. As it was not necessary to form a conventional image of the star, the mirrors did not need to be figured with conventional optical precision. To keep down the cost they were all ground to the same nominal focal length of 11 m with a tolerance of ± 15 cm. Again, in the interests of economy, the optical quality of the mirrors was specified so that their imperfections were roughly equal to the geometrical aberrations of the whole system. A programme of ray-tracing, taking the mirrors to be perfect, showed that the minimum circle of confusion was 10·85 m from the pole of the reflector and all the light was then contained in a circle 8 mm in diameter. The manufacturing tolerance on each mirror was therefore specified so that at their nominal focal length of 11 m all the light from a distant point source passed through a circle 1 cm in diameter and in the factory each mirror was tested and marked with its measured focal length. The mirrors were then mounted so that their actual focal lengths corresponded as closely as possible to the theoretical focal length for each position on the reflector. Thus, by making use of the spread of focal lengths in manufacture, a close approximation to a true paraboloidal mirror was obtained.

The mirrors themselves were front-aluminized and coated with silicon dioxide. It is worth noting that, even after ten years' work at Narrabri, they were still in good condition. We washed them at intervals of about six months with distilled water to which was added a mild detergent. Each mirror was mounted on a three-point spring suspension and its orientation could be adjusted from the back. We cemented an electrical heating pad to the back of each mirror and experience showed that about 12 W per mirror was sufficient to keep them free from dew and ice.

The optical system at the focus was mounted on a steel tube about 11 m long which projected from the centre of the reflector. This tube was guyed by stainless-steel rods attached to a simple framework of steel beams not directly coupled to the framework of the bowl supporting the mirrors. On each reflector there were two optical systems in separated boxes. One contained the main phototube and the other an auxiliary star-guiding phototube. The optical system of the main

Fig. 8 2.   One reflector of the stellar interferometer.

H

phototube is illustrated in fig. 8.3. The converging beam from the reflector was collimated by a negative lens with a diameter of 9 cm and a focal length of about 10 cm and was then passed through an interference filter 9 cm in diameter. In order to maximize the correlation the two optical filters were matched and uniform over their surfaces. Furthermore, they had a high transmission in the pass-band and a very low transmission outside it and, in order to maximize the spectral density factor, σ in equation (5.6), their pass-bands were as rectangular as possible. The original filters used at Narrabri had a central wavelength of 438·5 nm with a pass-band of $\pm 4·0$ nm. However, this wavelength proved to be too close to the Hγ (434·0 nm) line of the Balmer series, which is very wide in some A stars. Finally a wavelength of 443·0 nm was chosen; although this coincides with an interstellar absorption line, we were concerned with bright stars in which the feature is weak. The transmission curves of the two filters used

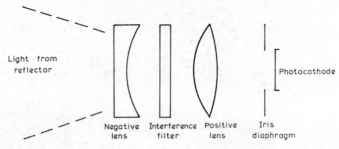

Fig. 8.3. The optical system.

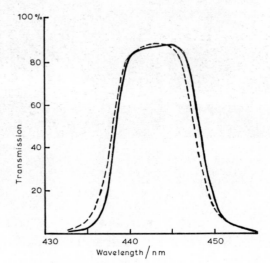

Fig. 8.4. Transmission versus wavelength of the two interference filters.

98

in most of the observations are shown in fig. 8.4. Their transmission at all wavelengths in the range of 400–700 nm outside the pass-band, never exceeded 0·1 per cent.

After passing the filter the light was focused by a positive lens with a focal length of 10·7 cm, through an adjustable iris diaphragm, onto the cathode of a photomultiplier. The photocathode of the most recent phototubes used at Narrabri had a maximum useful diameter of about 45 mm. The size of the 'image' on this cathode and the field of view are discussed later.

Fig. 8.5. Rear view of reflector being driven out of the garage by Dr. John Davis.

The optical components were mounted at the focus in a light-tight fibre-glass box coated with copper to form a shield against radio interference. The phototube was enclosed by a Mumetal screen to shield it magnetically. The front of this box was closed by a shutter remotely operated from the control desk. There was also a small pea-lamp mounted inside the box so that the photocathode could be illuminated when the shutter was closed. A bad feature of the mounting was that the central steel tube obscured some of the reflector as seen from the photocathode of the main phototube; the light from about 14 mirrors in the mosaic was lost in this way.

8.3 *Guiding and Control*

The movements of the two reflectors were controlled by an analogue computer which, given the sidereal time, the declination and right ascension of the star, and the latitude of the Observatory, calculated the azimuth and elevation of the chosen star. To follow the star in azimuth the reflectors moved around the track, and to follow it in elevation they tilted about a horizontal axis. At all times the line joining the two reflectors, which we shall call the baseline, was held at right-angles to the direction of the star; this was essential, not only to preserve a constant resolving power, but also to ensure that the light reached the two reflectors simultaneously. The length of the baseline could be varied from a minimum value of about 10 m to a maximum of 188 m. To make the two reflectors look in the same direction they were rotated on their turntables through half the angle subtended by the baseline at the centre of the track. To follow a star which transits north of the zenith the reflectors looked outwards from the centre of the track, and to follow a star south of the zenith, they looked inwards. This arrangement was necessary because the garage obstructed the extreme southern sector of the track.

On entering and leaving the garage the reflectors were controlled manually from a console on each tender. When they were clear of the garage a rail-operated switch allowed them to be controlled from the main desk in the control building.

The computed values of azimuth and elevation had an r.m.s. error of about $\pm 2 \cdot 5$ minutes of arc. However, these errors were usually small compared with the uncertainty in the pointing of the reflectors due to irregularities in the track which could introduce random errors of up to 15 minutes of arc. The combined effect of both these errors was removed by the use of an automatic photoelectric star-guiding system which employed a second phototube mounted at the focus of each reflector. This auxiliary phototube, which made use of one mirror of the mosaic, viewed the star through a rotating shutter and provided error signals corresponding to the azimuth and elevation of the star with respect to the optical axis of the reflector. These error signals were then used to correct the elevation and turntable angles transmitted by the computer to the appropriate reflector. It must be noted that the azimuth corrections were applied to the turntable motion and not to the position of the reflector on the track. Apart from considerations of dynamical stability, it was essential that the corrections should not alter the positions of the reflectors on the track since the baseline must always be constant in length and normal to the direction

Fig. 8.6. The two reflectors of the stellar interferometer in 1972.

Fig. 8.7. The control desk of the stellar interferometer. One reflector can be seen through the window.

of the star. Although the irregularities in the track disturbed the pointing of the reflectors, they did not significantly alter the length and orientation of the baseline.

The performance of the star-guiding system was monitored by a television camera at the focus of each reflector, and we found that the pointing accuracy depended on the speed of the trucks and on the local condition of the track. The maximum error was about ± 3 minutes of arc and the r.m.s. error was close to ± 1 minute of arc; in practice, once the reflectors were 'locked' they tracked the star without any further attention. Finally it is interesting to note that, because the star-guiding phototubes were mounted at the foci together with the main phototubes, they automatically compensated for the pointing errors due to sag of the long focal poles.

8.4 *The Phototubes*

To be suitable for use in a stellar interferometer a phototube must have a photocathode large enough to accommodate the 'image' of a star. As mentioned later, the 'image' in this case was roughly 25×25 mm and could be accommodated on a standard photocathode with an effective size of about 45 mm. Given an adequate size of photocathode, the most important parameters are the quantum efficiency (α), the excess noise $(\mu - 1)/\mu$, and the bandwidth or rise-time. From equation (5.17) it can be seen that the signal-to-noise ratio varies directly as the quantum efficiency and excess noise, but only as the square root of the bandwidth $(\Delta f)^{1/2}$ of the whole electronic system. The first phototubes used at Narrabri (R.C.A. type 7046) had flat semi-transparent antimony–caesium cathodes with a useful diameter of about 11 cm, a quantum efficiency of about 13 per cent at 440 nm and 14 stages of multiplication with an effective bandwidth (3 dB) of about 40 MHz. During the subsequent years we kept in close touch with the development of phototubes and introduced improved types as soon as we could get hold of them. Altogether five different types of phototube were used in ten years. The last type (R.C.A. Type 8850), installed in 1970, had a semi-transparent potassium–antimony–caesium cathode with a useful diameter of about 45 mm, a quantum efficiency of about 25 per cent at 440 nm and 12 stages of multiplication. The first dynode of these tubes was coated with gallium phosphide and had a very high gain (≈ 30), thereby reducing the excess noise in the multiplier chain and improving the signal/noise ratio by about 20 per cent compared with previous photo-tubes; under working conditions their effective bandwidth (3 dB) was about 60 MHz.

As a consequence of this continuous improvement the signal/noise ratio of the interferometer, on a given star, increased by about three times since it was first used to measure Vega in 1963. The majority

102

of this increase came from the steady improvement of quantum efficiency which contributed a factor of about 2; additional smaller factors came from improvements in the electrical bandwidth and excess noise.

Before installation on the reflectors the phototubes were tested in the laboratory. Their focusing controls were preset and their rise-times measured using nanosecond light flashes from a gallium–arsenide junction diode. The overall delay through the tube was also measured as a function of the supply voltages. As far as possible the supply voltages were kept constant for all stars and the gains of the phototubes were adjusted for different stars by pre-set controls which altered the voltages on either one or two dynodes in the multiplier chain in such a way that, to a first approximation, the overall delay was independent of the gain. Because the gain of all phototubes was really too high for this application, we had to run all tubes well below their rated voltage, which was undesirable because it reduced their bandwidth. The gain of each tube was individually adjusted for each star so that the r.m.s. output voltage on the cable to the correlator was roughly the same (≈ 1 mV), provided only that the associated anode current was less than about $100\ \mu$A. In the case of the brightest stars Sirius and Canopus, it was necessary to exceed this limit and to run the tubes with anode currents of up to $200\ \mu$A. Experience showed that, under these conditions, their gain slowly decreased with time.

8.5 The Correlator

At the output of each photomultiplier the fluctuating and steady components of the anode current were separated by a simple resistance-capacity filter. The steady component was connected directly to an anode-current integrator in the correlator. The fluctuating component was connected to the correlator by a low-loss coaxial cable, directly from one reflector and via a phase-reversing switch from the other. These coaxial cables were heavily screened; a cable with a solid aluminium outer conductor was used for the catenary, a double-screened flexible cable was used on the reflectors and at the central tower the signals were connected to the correlator by double-screened cables running in copper pipes.

A simplified outline of the correlator is shown in fig. 8.8. Following the earlier design described in § 6.1 we used double phase-switching. The phase of the signal from one phototube was inverted 10 000 times per second by a 5 kHz phase-switch mounted close to the input of the correlator; the phase of the other signal was inverted ten times per second by a phase-switch mounted close to the focus of the reflector. The 10 s switch was mounted in this way so that any radio signals picked up on the long cables to the correlator were not phase-switched and therefore did not produce spurious correlation. After phase-switching the signals were amplified by wide-band amplifiers, whose

103

pass-band was limited to roughly 20–100 MHz; frequencies below 20 MHz were severely attenuated to reject the majority of radio signals, and between 20 and 100 MHz the filter networks produced a rising characteristic to compensate for the falling response of the long cables and of the phototubes. The overall bandwidth between 3 dB points of the whole system, including the phototubes, was about 55 MHz.

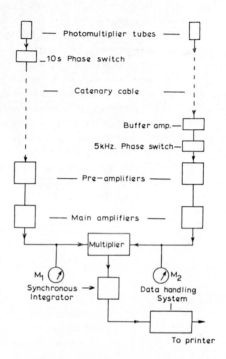

Fig. 8.8. Outline of the correlator.

The outputs of the main amplifiers were applied to a linear multiplier which proved to be the most difficult single component to develop. If the input voltages to the multiplier are e_1, e_2, then one major problem is to reduce unwanted components such as $e_1^2, e_2^2, e_1e_2^3, e_2e_1^3$, etc., to negligible values compared with the desired product e_1e_2 in the output. These spurious components must be reduced to very low levels; they are undesirable, not so much because they increase the noise but because they tend to produce spurious correlation through various second-order processes which in turn produce drifts in the zero of the correlator output. These drifts will be discussed later. In the early stages of the work at Narrabri a variety of multipliers was developed, using balanced arrangements of two or four vacuum tubes, but none

of them was satisfactory.   The final design was based on a multiplier
originally developed by Frater (1964), and the circuit and performance
have been discussed by Allen and Frater (1970).   Since it is such an
important component of an intensity interferometer and has proved so
difficult to develop, the circuit is shown in fig. 8.9.   The basic multiplier consists of the four transistors T 1, 2, 3, 4.   One input voltage
(No. 1) from the main amplifier is applied via T 1, 2, to the emitters
of the balanced pair T 3, 4.   The other input (No. 2) is applied
directly to the base of T 4.   In his paper Frater shows that the collec-
tor currents of T 3, 4, contain, in push–pull, a component linearly
proportional to the product of the two inputs.   Transistors T 5, 6 ,7,
convert this push–pull output into a single-ended output suitable to
the synchronous integrator which followed the multiplier.   Allen and
Frater found that the multiplier was linear over a wide range of input
voltages and that the levels of the unwanted components in the output
were acceptably low; experience at Narrabri showed that it was
adequately stable and easy to adjust.

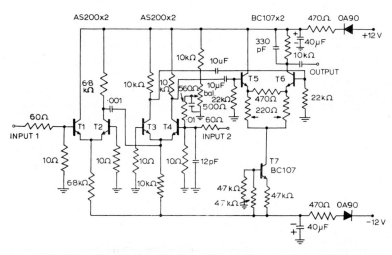

Fig. 8.9.   The linear multiplier.   From Allen and Frater (1970).

In the output of the multiplier the presence of correlated signals
from the two phototubes produced a 5 kHz square wave whose phase
reversed every 10 s.   This wave was submerged in noise and, to reduce
the noise, the bandwidth of the multiplier output was limited to 50 kHz
which is the minimum value necessary to pass all the significant
harmonics of the 5 kHz square wave.   Even so, the r.m.s. level of the
square wave was always less than 1 per cent of the noise in the multiplier
output.   The selective amplification of this square-wave was carried

105

out by the synchronous integrator illustrated in fig. 8.10. The two transistors T 3, 4, were switched at 5 kHz, synchronously with the phase-switch, so that alternate half-cycles of the wave were stored in the capacitors C_1 and C_2. In effect, this circuit acts as a comb filter tuned to the fundamental and odd harmonics of the 5 kHz signal. Frater (1965) shows that the effective bandwidth at each of these frequencies is $1/2\pi R(C_1 + C_2)$ and, in the present case, was 8 Hz. Thus, in the absence of correlation, the output of the synchronous integrator was the sum of the random noise in five 8 Hz bands centred on 5, 15, 25, 35 and 45 kHz; if a correlated signal was present it appeared as a 5 kHz 'square' wave whose phase reversed every 10 s.

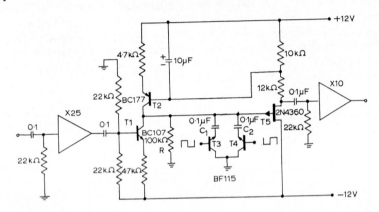

Fig. 8.10. The synchronous integrator. From Allen and Frater (1970).

Following the synchronous integrator there was a conventional synchronous rectifier at 5 kHz. The output of the rectifier was then integrated for 10 s, corresponding to one position of the 10 s phase-switch, in a very precise linear integrator.

Finally, the data-handling system, following the practice of the earlier correlator, acted as a synchronous rectifier for the 10 s phase-switch and then integrated the result. As before, this system had the advantage that the final stage of synchronous rectification was numerical and did not contribute to the zero-drift of the correlator output. It worked as follows: the output of the 10 s integrator, following the synchronous integrator, was read every 10 s by a digital voltmeter and displayed together with its sign; the integrator was then re-set to zero, the 10 s phase-switch operated and the cycle repeated. A sign-sensing unit examined the sign of the digital voltmeter reading and if it was positive, it was accumulated in a number store S_1; if it was negative, it was put in a number store S_2. In the next 10 s cycle the role of these two stores was reversed. After ten such periods of integration, corresponding to a total time of 100 s, the contents of the two number stores S_1 and S_2 were read out, decoded from binary to serial decimal form and printed. The printer then took the difference in a sub-total register which was also printed every 100 s; this number was the sum, or long-term integration, of the correlator output.

In addition, the anode currents of the two phototubes were also averaged over the whole 100 s period by linear integrators and were printed. Thus, every 100 s the printer printed a line of numbers—the two numbers in the stores S_1, S_2, their accumulated difference, and the anode currents of the two phototubes.

As we have already noted in §6.1, the principal problem in designing the correlator was to eliminate very small but irregular drifts in the zero level of the output. In practice it proved very difficult to identify the sources of the drifts because they were so small and variable and it took so long to make each measurement. The main sources were (1) coupling between the inputs to the correlator, (2) external radio signals, (3) amplitude modulation of the input signals by the phase-switches, and (4) effects due to the asymmetrical amplitude distribution of the signals from the phototube. A simple calculation shows that, to avoid spurious correlation due to coupling between the inputs, an isolation of about 120 dB is required; this was achieved by the use of heavily screened cables. The second effect, correlation produced by external radio signals, was serious in the original installation at Narrabri but was eliminated partly by screening the cables and partly by rejecting frequencies below 20 MHz. The third effect, spurious correlation due to unwanted amplitude modulation in the phase-switches, is discussed in some detail by Allen and Frater (1970), who show how this correlation is produced by non-linearities in the multi-plier. This particular trouble was reduced to acceptable levels by the use of an adequately linear multiplier and by reducing amplitude modulation by the phase-switches to a low level. At Narrabri the phase-switches, due to difference in gain in their two positions, intro-duced less than 1 per cent of amplitude modulation over the whole band-width of the correlator. Lastly, there were the effects produced by the asymmetrical nature of the input 'noise'; thus, for a star of magnitude $+2{\cdot}5$, the number of photoelectrons emitted per second from the photocathodes was only about 2×10^8 and each gave rise to a pulse of duration about 10 ns. It follows that the output signal from the phototubes consisted of superimposed pulses and that, on the average, there were only two pulses at any given moment. Under these conditions the amplitude distribution was markedly asymmetrical, all the larger peaks being in one direction. These peaks swung the amplifiers and the multipliers over a wide range of their characteristics, first in one direction and then in the other as the phase-switch reversed. Non-linearities in the amplifiers and the multiplier converted these phase-reversals into modulation products at the phase-switching

107

frequencies and produced small spurious correlations which tended to vary with the signal level and precise balance, and hence temperature, of the multiplier. These effects were reduced and stabilized, but not completely eliminated, by the use of the linear multiplier described above and by keeping the input voltages from the phototubes within the same fairly narrow range of levels for all observations.

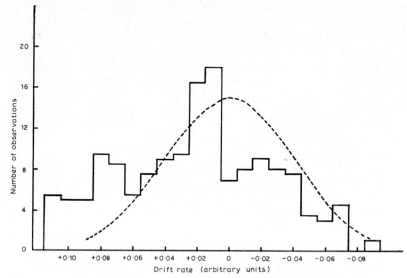

Fig. 8.11. The zero-drift of the correlator at Narrabri. The full line is a histogram of the correlation measured in runs of 12 hours' duration on uncorrelated light sources; the broken line is the expected distribution for a perfect correlator with no zero-drift. From Allen and Frater (1970).

To increase the stability of the correlator and in particular to stabilize any residual zero-drifts, the whole equipment was totally enclosed in a metal cabinet and cooled with air at $72 \pm 2°$ F. All supply voltages were well stabilized and the equipment was run continuously day and night throughout the observing programme. When the correlator was not measuring a star, dummy measurements were made with the phototubes illuminated by small lamps giving the the same light flux as the star. Under these conditions the correlator should, of course, register a total correlation of zero with a dispersion equal to the statistical uncertainty due to the noise in the correlator output. The solid line in fig. 8.11 shows the total correlation measured on dummy runs of 12 hours' duration during the first six months of 1969. The broken line shows the expected distribution of measurements for a correlator with no drift but with the same output noise. It can be seen that, although the correlator was not perfect, the mean drift over 12 hours was less than the statistical uncertainty in the output

108

due to noise.   It should be remembered that the effect of any mean drift was reduced by the observing procedure in which the output of the correlator was observed every day in a dummy run lasting at least 12 hours; the appropriate three-day running mean of this output was then subtracted from the correlation observed from the star.

There is one further fact about the correlator which is worth recording.   About 5 per cent of the noise in the output was due to sources within the correlator and not to the input signals.   The majority of this noise was generated in the first stage of the amplifiers.

The first description of the correlator at Narrabri was given by Browne (Hanbury Brown and Browne, 1966) and a more recent one by Allen and Frater (1970).

8.6   *An Alternative Design of Correlator*

In the early stages of the work at Narrabri the difficulties encountered in stabilizing the zero level of the correlator output suggested that it would be worth while to explore an alternative design.   Since most of the trouble appeared to be due to unwanted amplitude modulation of the signal by the phase-switches and to the asymmetric character of the signal itself, it was decided to build an amplitude-limiting correlator in which the signals were passed through a bi-directional amplitude limiter before multiplication.   In this way it was hoped to remove all the effects due to the amplitude of the signals and to preserve only their relative phase.   It is, of course, to be expected that such a non-linear operation will cause loss of signal/noise ratio since some of the information is carried by the amplitude fluctuations.   However, if the overall signal/noise of the correlator is limited by irregular drifts in the output then some loss can be tolerated in exchange for improved stability.

As a first step a theoretical investigation of the effects of amplitude limiting was carried out by Yerbury (1967).   He calculated the loss of signal/noise ratio, relative to a completely linear correlator, as a function of the limiter level for the case where only one signal is limited and also for the case where both are limited.   He defined the loss factor as F, where $F = 10 \log_{10}[(S/N)_1/(S/N)_0]$ decibels and $(S/N)_1$, $(S/N)_0$ represent the output signal/noise ratios of the amplitude-limiting correlator and the linear correlator respectively.   In the case where only one signal is limited he found $F = -0.525$ dB, corresponding to a loss in signal/noise of about 11 per cent; this result assumes that the signal is limited at a level very small compared with its r.m.s. value.   In the case where both signals are limited to this low level, he found $F = -1.081$ dB, corresponding to a loss of signal/noise ratio of 22 per cent.   This loss decreases rapidly as the limiting level is raised but so, presumably, do the benefits of the limiter.

As a second step Yerbury (1968) built an amplitude-limiting correlator and compared it directly with the linear correlator at Narrabri which we have just described.   He first measured the signal/noise

109

ratio of both correlators on an 'artificially' correlated signal produced by cross-coupling the two phototubes of the interferometer through an attenuator network. He found the loss of signal/noise ratio in the amplitude limiting correlator, relative to the linear correlator, to be $1 \cdot 30 \pm 0 \cdot 29$ dB; the theoretical value, based on the actual limiting level used in these experiments, was $0 \cdot 90$ dB. He then repeated this comparison under more realistic conditions by observing the bright star α Eridani with both correlators. This test confirmed that there was a loss of signal/noise ratio in the amplitude-limiting correlator of $1 \cdot 72 \pm 0 \cdot 46$ dB.

The output zero level of the amplitude limiting correlator was found to be almost completely free of drift, provided that the signal was limited at a level small compared with its r.m.s. value. This conclusion was based on a few runs of 13 hours' duration with different limiter levels. To establish the zero-drift with certainty would have required tests lasting several months. However, these were not carried out because, by the time the amplitude correlator was built, the performance of the original linear correlator had been improved to the point where its zero drift was no longer a significant limitation to the signal/noise ratio. It was therefore decided that the loss of roughly $1 \cdot 5$ dB in signal/noise, or about 30 per cent, was too high a price to pay for the introduction of an amplitude-limiting correlator and the new correlator was not used.

In conclusion, this work confirmed that an extremely stable correlator can be built using amplitude limitation. It is an attractive technique provided that the small loss in signal/noise can be afforded.

how the observations at Narrabri were made

9.1 *Focusing the Reflectors*

The reflectors of the interferometer were ready for their first test in October 1962. As a first step, a small telescope was rigidly fixed to one end of the axle supporting the mirror framework in order to define permanently a line parallel to the optical axis. Two lights were mounted on a horizontal bar on a tree about one mile away so that they could be seen by the reflector when it was parked in the garage; the other reflector was temporarily removed. The horizontal separation of the lights was equal to the distance between the fixed telescope and the centre of the reflector. The reflector was pointed so that one light was central in the telescope and the reflector itself was then 'focused' on the other light as follows: a mobile tower was positioned so that an observer could see the image of the distant light on a ground-glass screen at the focus, the distance of this screen from the true focus being adjusted to allow for the finite distance of the light. The individual mirrors were then aligned to give the minimum possible size of image on the screen which proved to be a roughly circular patch 13 mm in diameter corresponding to an angular field of about 4 minutes of arc.

Following this adjustment the screen was moved to the correct distance from the reflector for sources at infinity, and a remotely operated 35 mm camera was mounted so that pictures of the image could be taken under working conditions. The reflector was pointed at Jupiter and the image was photographed over a wide range of elevations. To our disappointment the size and shape of the image varied greatly with elevation; at 70° elevation it deformed into an ellipse measuring 60 × 25 mm. A thorough investigation of the trouble was made and showed that the weight of the reflector was bending the main steel tube carrying the framework; the maximum deflection of this tube was 3·5 mm, which was sufficient to distort the reflector framework and enlarge the image.

It was not easy to modify the structure at Narrabri and after some discussion we decided to accept the larger image and to evolve an optimum method of focusing the reflectors. We first measured the movements of the individual mirrors with a television camera mounted at the focus and, using these data, we evolved a system of aligning the mirrors into a predetermined pattern on the distant light. This pattern was designed so that the mirrors produced a minimum size of image at about 50° elevation and over most of the working range of elevation, the image was roughly 25 × 25 mm. This size could be accommodated on the photocathode which had a diameter of 45 mm and corresponded to an angular resolution in the sky of about 8 × 8 minutes of arc; the angular field of view of the whole photocathode was about 15 minutes of arc. Experience showed that this alignment of the mirrors had to be repeated at intervals of about six months.

The star-guiding system depended on only one mirror in the mosaic and this was aligned so that when the star-guiding system was 'locked-on' a star the centre of the main image from the reflector fell on to the centre of the cathode. An initial adjustment was made when the reflectors were aligned on the distant light, but a more precise adjustment was made with both reflectors tracking a star.

9.2 *Equalizing the Time Delays*

Another important adjustment which had to be made before every sequence of observations was to equalize the electrical time delays in both arms of the interferometer. As already noted in §8.4, the time delays in the phototubes were measured, as a function of voltage, in the laboratory with pulsed light. Once the working voltages were chosen, any difference between the time delays in the two phototubes was compensated by an extra short length of cable in the appropriate output. The electrical lengths of the long cables from the phototubes to the correlator were equalized by disconnecting the cables from the phototubes and connecting them to a standard source of wide-band noise—a saturated diode—mounted on the wall of the garage. The correlation due to this standard source was measured as a function of the length of small pieces of cable added to one or other of the input cables to the correlator. When the correlation was a maximum the electrical lengths were equal and in practice it was simple to equalize them to better than 5 cm in this way.

As a final and important check on the whole system a bright star was observed with minimum baseline and the normalized correlation measured as a function of small lengths of cable added first to one cable and then to the other. A typical 'delay check' using Sirius is shown in fig. 9.1.

9.3 *Measuring the Zero-drift, Gain and Noise Level of the Correlator*

The correlator was kept running continuously, 24 hours a day, during an observing programme. When not observing a star the light-tight shutters were closed and the phototubes illuminated by small lamps. The light from these lamps being uncorrelated, the output of the correlator became a measure of spurious correlation or zero-drift. For each period of 24 hours, measured from noon to

112

noon, the total correlation for the dummy runs was recorded as a measurement of the drift D; the r.m.s. uncertainty in this drift ($\pm \sigma_D$) was calculated from the duration of the dummy runs and the r.m.s. noise (σ_{OBS} see later) in the correlator output. These data were used as a check on the correlator and to correct the stellar observations for drift as described in § 10.1.

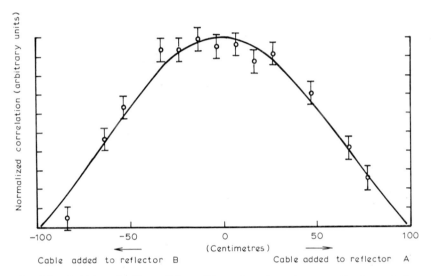

Fig. 9.1. A 'delay check' on Sirius. The points show the normalized correlation observed with different small additional lengths of cable inserted in the cables to each reflector.

The gain of the correlator was measured directly before and after every night's programme on a star. The cables were disconnected from the phototubes at the focus of each reflector and connected to the standard noise generator in the garage. The correlation was recorded for several minutes so that the uncertainty in the measurement of gain was less than 1 per cent. The overall gain of the system depended not only on the correlator but also on the loss in the cables, which varied with the outdoor temperature. There was a change of roughly 5 per cent in the gain for a change of 20 K. The temperature was therefore recorded when the gain was measured and at half-hourly intervals throughout the night. The gain for any particular observation during the night was found by linear interpolation.

The noise level at the output of the correlator was monitored every day. The r.m.s. fluctuation in about 100 'cycles' (each cycle corresponding to an observation of 100 s) was computed and used to find σ_{OBS} the r.m.s. uncertainty in a single cycle. The average anode currents of the phototubes over the same period were also computed

113

from the print-out. These data were then used to find a normalized or standard uncertainty σ_{STD}. It is simple to show that

$$\sigma_{OBS} \propto G_c (G_{p1} G_{p2} \langle i_1 i_2 \rangle)^{1/2} \tag{9.1}$$

where G_c is the gain of the correlator, G_{p1}, G_{p2} are the gains of the phototubes and i_1, i_2 are the phototube anode currents. Hence we can define a standard uncertainty σ_{STD} which is independent of the light fluxes and the gain of the correlator but not of the phototubes, as

$$\sigma_{STD} = \sigma_{OBS} / (\langle i_1 i_2 \rangle^{1/2} G_c). \tag{9.2}$$

This standard uncertainty was computed for each day and was used as a check on the correlator and also in the analysis of the observations. In practice the value of σ_{STD} was usually averaged over a few days and used to compute the uncertainty σ_c in any observation of duration n cycles, where

$$\sigma_c = \sigma_{STD} G_c \langle i_1 i_2 \rangle^{1/2} / n^{1/2} \tag{9.3}$$

and i_1, i_2 are the appropriate anode currents of the two phototubes and G_c is the correlator gain.

9.4 Choosing the Baselines and Exposure Times

Two important factors governing the total time spent in measuring a star for a given accuracy in the result are the choice of the baselines and exposure times. At Narrabri this choice was particularly critical because the exposure times were so long.

If we assume that the star is single, then the angular diameter of the equivalent uniform disc can be found from measurements at only two baselines; furthermore, it can be shown that the use of only two baselines yields the highest precision in the final result for a given total exposure. Ideally one of these baselines should be zero which, with the Narrabri instrument, led to a choice of the minimum possible baseline ($d_1 = 10$ m) as the first baseline for all stars. The optimum length of the second baseline (d_2), together with the ratio of exposures (T_2/T_1) at the two baselines, was calculated for an estimated angular diameter (θ) of the star. As an example, fig. 9.2 shows how the uncertainty σ_θ in the final measured angular size varies with d_2 and T_2/T_1 for a given total exposure time. These particular curves were calculated for the case where $\pi d_1 \theta / \lambda = 0 \cdot 5$, which corresponds to the representative case of a first magnitude main sequence star observed with a first baseline of 10 m ($d_1 = 10$ m, $\lambda = 443 \cdot 0$ nm, $\theta = 1 \cdot 5 \times 10^{-3}$ seconds of arc). It can be seen that in this particular case, the most precise measurement of θ is to be expected when $T_2/T_1 = 4$ and the second baseline has a length given by $\pi d_2 \theta / \lambda = 2 \cdot 25$.

114

In practice such calculations were always used to find the length of the second baseline but were only used as a rough guide to the ratio of the observing times, as in choosing these times there were other considerations to be taken into account. For example, to minimize the effects of possible systematic errors with elevation angle, every effort was made to observe stars over exactly the same range of elevations at each baseline; furthermore, to reduce any systematic errors which might vary slowly with time, observations at the two baselines were interleaved on successive nights as far as possible. These procedures usually led to a ratio $T_2/T_1 \approx 2$ which is significantly less than the theoretical optimum. Inspection of fig. 9.2 shows that the corresponding theoretical loss of precision is only a few per cent.

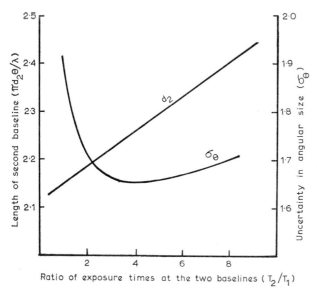

Fig. 9.2. The optimum baselines and exposure times for measurements at two baselines. The curve (σ_θ) shows the uncertainty in measurements of (θ) the angular size versus (T_2/T_1) the ratio of exposures at the two baselines. The line (d_2) shows the corresponding optimum length of the second baseline. The curves are calculated for the shortest first baseline possible at Narrabri, $\pi d_1 \theta/\lambda = 0.5$.

9.5   Observational Procedure

Immediately before an observing programme the gain of the correlator was calibrated. The two reflectors were then driven out of the garage by hand using the controls on each tender. When they were at a safe distance from the garage an automatic interlock allowed control of their movements to be transferred to the desk in the central control building. They were then driven by remote control to approximately

115

the correct positions for starting the observations and a check made by calibrated marks on the track that the actual positions of the reflectors corresponded to the positions indicated on the control desk. They were then set manually from the control desk so that their positions in three coordinates—track position, turntable and elevation angle— were within a few minutes of arc of the position demanded by the computer. Control was then switched from manual to automatic and the reflectors were completely controlled by the computer. If everything was right the reflectors should by then be pointing at the star, but in practice there were always pointing errors of several minutes of arc due to imperfections in the track. To correct these errors the star-guiding system was switched on and the direction of the star relative to the optical axis of the reflectors was indicated by bright spots of light on two cathode-ray tubes on the control desk. These spots were centralized manually by controls which corrected the turntable and elevation angles transmitted to each reflector; the star-guiding system was then switched from manual to automatic and kept both reflectors pointing at the star with an r.m.s. error of about 1 minute of arc.

When both reflectors were 'locked-on' a star, the small lamps in front of each phototube were switched off, the shutter opened and the run on the star started. Every half hour throughout the night readings were taken of the computed azimuth and elevation of the star and compared with a local ephemeris for the star as a check on the computer. At the same time we recorded the pointing corrections made by the star-guiding system, the anode currents of both phototubes, the voltage levels in the correlator, the temperature and the wind speed. At various times during the night the contribution of the light from the night-sky to the total light flux was measured by pointing the reflectors about half a degree away from the star for two or three cycles of the correlator.

As a quick visual check on the behaviour of the whole system a graphical record was always made of the cumulative correlation, showing it as a function of the number of 100 s cycles. An example is given in fig. 9.3. It is plotted for a period of 24 hours from noon to noon. The first part from cycle 0 to cycle 160 is a dummy run where no correlation is expected; the second part from cycle 200 to cycle 500 is an observation of the star β Crucis at two baselines and the increase in the cumulative correlation can be seen clearly; the third part from cycle 550 to cycle 750 is a continuation of the dummy run.

When the run on the star was completed (usually about one hour before dawn) the shutters were closed and the reflectors put back in the garage. The gain of the correlator was measured and the small lamps adjusted to give phototube anode currents equal to the average observed on the star.

Observations were not usually made on stars below 30° elevation nor when poor atmospheric transmission reduced their light flux below

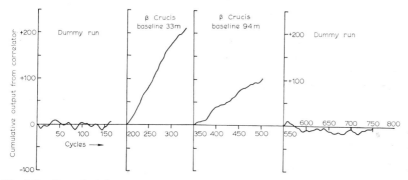

Fig. 9.3.   Record of the cumulative correlation for 24-hour period (20–21 May 1965).

about 70 per cent of its normal value, nor when the contribution of moonlight to the total flux exceeded about 10 per cent of the total. Every effort was also made not to expose the reflectors to high wind, rain or hail.   If the wind exceeded about 25 knots the reflectors were returned to the garage; high winds reduced their pointing accuracy and could have upset the alignment of the mirrors; furthermore, one was always anxious that the wind might get even stronger and do some structural damage.   While it was not really necessary to avoid rain, there was always the possibility that it might turn to hail, which at Narrabri can flatten a wheat crop in a few minutes and might be expected to damage the front-aluminized mirrors.   Experience of the past ten years shows that observations were made on about 60 per cent of all nights on which they were scheduled.

CHAPTER 10

analysing the data

10.1   *The Normalized Correlation*

The first step in analysing the data for a single night was to correct the observed correlation for zero-drift. A three-day running mean of the zero-drift ($\bar{D} \pm \sigma_D$) centred on the date of observation was computed from the daily measurements of drift (see § 9.3) and subtracted from the average correlation observed per cycle $\overline{c(d)}$ to give the true average correlation $\overline{c_0(d)}$ for that night, so that

$$\overline{c_0(d)} = \overline{c(d)} - \bar{D}. \tag{10.1}$$

This procedure was based on two assumptions; first, it was assumed that the drift during a dummy run with constant light flux was the same as that when observing a star with varying light flux; secondly, the drift was assumed to be steady over periods of three days. We made every effort to ensure that both these assumptions were justified. For example, the small lamps in front of the phototubes were adjusted for each dummy run so that the average phototube currents were equal to those actually observed on the star; furthermore, the correlator was run continuously under constant conditions of supply voltage and ambient temperature. The zero drift itself was checked every day and was normally small but not negligible; it seldom exceeded about 5 per cent of the zero-baseline correlation from a first magnitude star. If it was found to be abnormally high, all the associated results were rejected and attention was given to the correlator. All the available evidence showed that this procedure was successful and that if there were any systematic errors in the final results due to zero-drift, they were negligibly small.

The next step was to subtract from the total observed photomultiplier and anode currents $i_1(T), i_2(T)$ the contribution of the night-sky, to find the currents due to the star alone $i_1(S), i_2(S)$.

Now it follows from equation (5.1) that the correlation $c_0(d)$ in any given cycle varies with current and gain so that

$$c_0(d) \propto i_1(S) i_2(S) G_C \tag{10.2}$$

and the uncertainty $\pm \sigma_C$ in this correlation varies as

$$\sigma_C \propto (i_1(T) i_2(T))^{1/2} G_C \sigma_{STD} \tag{10.3}$$

where G_C is the gain of the correlator and σ_{STD} is defined in equation (9.2). Hence $c_N(d)$ the correlation normalized to standard currents and gain is given by

$$c_N(d) = \frac{c_0(d)}{G_C i_1(S) i_2(S)} \pm \frac{(i_1(T) i_2(T))^{1/2} \sigma_{STD}}{G_C i_1(S) i_2(S)}. \tag{10.4}$$

Thus the weighted mean correlation for a single night of n cycles is given by

$$\overline{c_N(d)} = \frac{\sum_n [c_0(d) i_1(S) i_2(S)/i_1(T) i_2(T)]}{\sum_n [\bar{G}_C (i_1(S) i_2(S))^2/i_1(T) i_2(T)]} \tag{10.5}$$

where $\bar{G}_C$ is the mean gain of the correlator. If we now assume that $i_1(T) = i_1(S)$ and $i_2(T) = i_2(S)$, solely for the purpose of weighting the results, then we may write

$$\overline{c_N(d)} = \frac{\sum_n c_0(d)}{\sum_n \bar{G}_C i_1(S) i_2(S)} = \frac{\sum_n c_0(d)/n}{\sum_n \bar{G}_C i_1(S) i_2(S)/n} \tag{10.6}$$

which can be written

$$\overline{c_N(d)} = \overline{c_0(d)}/[\bar{G}_C . \overline{i_1(S) i_2(S)}]. \tag{10.7}$$

Thus the normalized weighted mean correlation for a run on a star is equal to the mean correlation for the run divided by the mean phototube currents due to the star and the mean gain of the correlator. In practice this fact greatly simplified the analysis of the results because it was not necessary to normalize and weight each 100 s period individually.

The statistical uncertainty ($\pm \sigma_N$), due to noise alone, in the mean normalized correlation for the whole run is given by

$$\sigma_N = [\sigma_D{}^2 + \sigma_C{}^2]^{1/2} \tag{10.8}$$

where σ_D is the uncertainty in the three-day running mean of the drift and is given by

$$\sigma_D = \bar{G}_C \sigma_{STD} (i_1 i_2)^{1/2}/p^{1/2} \tag{10.9}$$

where p is the number of 100 s cycles used in finding the drift and σ_C is the statistical uncertainty in the correlator output given by equation (10.3).

The final value of the weighted mean correlation for a given baseline $\overline{c_N(d)}$ was found by summing the results obtained on different nights.

The value for each night was weighted by the square of its uncertainty so that for the results taken on r nights

$$\overline{c_N(d)} = \frac{\sum_r \overline{c_N(d)}/\sigma_N{}^2}{\sum_r 1/\sigma_N{}^2} \pm \frac{1}{(\sum_r 1/\sigma_N{}^2)^{1/2}} \tag{10.10}$$

where the uncertainty is only that due to statistical fluctuations in the correlator output. There are, of course, other factors which contribute to the uncertainty of the final result and these will be considered in the following section.

10.2 *Estimating the Uncertainty in the Normalized Correlation*

The final uncertainty in the results was determined by two groups of factors—those that were invariant with baseline and those that were not.

The first group comprised the statistical noise in the correlator output, errors in the normalizing factors (phototube anode currents and correlator gain), the possible effects of scintillation, and the change of optical bandwidth with elevation angle. The uncertainty due to statistical noise is calculable and has been discussed in § 10.1. The errors in the phototube anode current product $(i_1(T)i_2(T))$ were a function of the linearity, calibration and zero level of the current integrators and introduced an uncertainty of less than 1 per cent into the results. The errors in the correlator gain were due to changes in the ambient temperature during the night; these changes altered the dielectric loss in the long cables from the phototubes to the correlator by as much as 5 per cent during the night. The gain of the system, including the cables, was measured before and after every run and the temperature was recorded; nevertheless, there was a residual uncertainty of about 1 per cent in $\bar{G}_C$ the mean gain of the correlator. The effects of scintillation are discussed in § 11.11 where it is shown that they were almost certainly negligible. The change of optical bandwidth with elevation angle was due to mechanical distortion of the reflectors which changed the shape and size of the 'image' of the star. There were corresponding changes in the angular dispersion of the light passing through the interference filters and hence small changes in the optical bandwidth B_0. From equation (5.1) it can be shown that $\overline{c_N(d)}$ is inversely proportional to the optical bandwidth and hence there was a small systematic variation of normalized correlation with elevation angle. Extensive measurements of bright stars showed that the normalized correlation did not vary significantly from $30°$ to $55°$ elevation and then decreased smoothly with a loss of about 10 per cent at an elevation of $75°$.

Although this first group of factors increased the uncertainty in the final result they did not introduce systematic errors into the *ratio* of the

120

normalized correlations at different baselines; hence they did not affect the measurements of angular diameter. This statement depends, of course, on the fact that observations at each baseline were carried out over the same range of hour angles and hence the same elevation angles, and the programme was always planned so that the dates of observations at different baselines were interleaved, thus minimizing any effects of slow changes in the equipment or the environment.

The second group of factors—those believed to vary with baseline—were likely to introduce systematic errors into the angular diameters and therefore very great care was taken to estimate their magnitude. This group comprised the possibility of spurious correlation due to Čerenkov light pulses from the night-sky and also to radio interference, the possibility of baseline-dependent coupling between the arms of the interferometer and the loss of correlation due to misalignment of the baseline. Upper limits to unwanted correlation due to Čerenkov light pulses, radio interference and coupling between the arms of the interferometer were established experimentally as described in § 11.10.

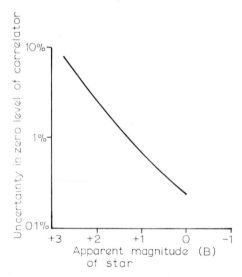

Fig. 10.1. The uncertainty in the zero-level of the correlator due to possible sources of unwanted correlation (see text). The uncertainty is expressed as a percentage of the zero-baseline correlation from the star and is plotted against the apparent magnitude of the star under observation.

The numerical results of these tests are shown in fig. 10.1 and represent an uncertainty in the zero level of the correlator and therefore an additional uncertainty in the normalized correlation. The loss of correlation due to misalignment of the baseline was caused by differential delays in the arrival time of the light at the two phototubes. Thus, taking the electrical bandwidth to be rectangular with a width of

121

100 MHz there was a loss of correlation of about 1 per cent (see equation (4.35) and fig. 4.7) for a differential delay of 0·33 ns. At the longest baseline used at Narrabri—150 m—this corresponded to a misalignment of the baseline of about ± 3 minutes of arc when observing a star at an elevation angle of 45°. Since in practice the azimuth was usually maintained with a precision of roughly ± 3 minutes of arc, any loss of correlation was only significant at long baselines and low angles of elevation. This loss was computed from the azimuth errors of the computer which were recorded every half hour and, if necessary, a correction was made to the measured correlation; the uncertainty in this correction was included in the uncertainty of the result.

Although the total uncertainty in the measured values of normalized correlation and in the derived values of zero-baseline correlation and angular diameter was estimated by taking into account all the factors outlined above, the uncertainty in the final result, except for the two brightest stars (α C Ma and α Car), was largely due to the statistical noise in the correlator output. It is interesting to note that, if a significant improvement were to be attempted in the signal/noise ratio of the instrument—for example by increasing the size of the reflectors —it would also be necessary to lower the limits set by some of the minor factors described above.

10.3 *Finding the Angular Diameter of Single Stars*

10.3.1 *The angular diameter of the equivalent uniform disc*

It is convenient to express the results for single stars as the angular diameter θ_{UD} of the equivalent circular disc of uniform surface brightness. This result may then be interpreted later (§ 10.3.2) in terms of particular models of the star which take into account limb-darkening. The values of the mean normalized correlation $\overline{c_N(d)}$ measured at each baseline, together with their uncertainties were calculated from equation (10.10). These results were then fitted to a theoretical curve in a computer. In the simple case, where the star is not significantly resolved by the individual reflectors ($\Delta \approx 1$), the shape of this curve is given by

$$\Gamma^2(\lambda_0, d) = [2J_1(x)/x]^2 \tag{10.11}$$

where $x = \pi d\theta_{UD}/\lambda_0$, λ_0 is the effective wavelength and θ_{UD} is the angular diameter of the equivalent uniform disc. The least-squares fitting of the data to this curve was carried out by an iterative programme. Reasonable initial values were taken for $\overline{c_N(0)}$ (the normalized correlation at zero baseline) and θ_{UD}, and corrections $d\overline{c_N(0)}$ and $d\theta_{UD}$, and corrections $d\overline{c_N(0)}$ and $d\theta_{UD}$ were computed to minimize the weighted squared differences between the observations and the theoretical curve. The r.m.s. uncertainties in $\overline{c_N(0)}$ and θ_{UD} were also computed. The

122

calculation was then repeated using corrected initial values and the iteration was continued until $d\theta_{UD}/\theta_{UD}$ and $\overline{dc_N(0)}/c_N(0)$ were $< 10^{-4}$.

If the angular diameter of the star is large enough to be partially resolved by the individual reflectors, the theoretical curve is modified as we have already seen in §5.3. The shape of the curve is then given by equation (5.12) and is illustrated in fig. 5.1. The broken line for the case $\Delta = 0\cdot90$ corresponds approximately to observations of Sirius with the reflectors at Narrabri. For most of the stars observed at Narrabri the effects of partial resolution were negligibly small but, where necessary, the values of $\overline{c_N(0)}$, θ_{UD} and Δ were found from the data by a least-squares fit to equation (5.12) using an iterative procedure as before. The results for three stars are shown in fig. 10.2. One star, β Crucis, has a faint companion and therefore has a lower value of zero-baseline correlation (§10.3.3).

10.3.2 The effects of limb-darkening

We saw in chapter 4 that in an intensity interferometer the correlation is proportional to the square of the amplitude of the Fourier transform of the intensity distribution across the stellar disc, in contrast to Michelson's interferometer in which the visibility of the fringes is *linearly* proportional. In consequence, unless the star is very bright, the details of the distribution are likely to be lost because they are carried by high-order Fourier components of low/signal noise ratio.

This point has been illustrated in a discussion of the effects of limb-darkening by Hanbury Brown and Twiss (1958 a). They calculated the variation of correlation with baseline for a star with a uniform disc of angular diameter θ_{UD}, and also for a limb-darkened star of angular diameter θ_{LD}. They took a simple approximate form of the conventional cosine law of limb-darkening in which the intensity at any angle θ from the centre of the star is given by

$$I_\lambda(\theta) = I_\lambda(0)[1 - u_\lambda\{1 - (1 - \theta^2/\theta_{LD}{}^2)^{1/2}\}] \qquad (10.12)$$

where θ_{LD} is the true angular diameter of the star and u_λ is the conventional limb-darkening coefficient. Their results for a completely limb-darkened star $(u_\lambda = 1)$ of diameter θ_{LD} are shown by the crosses in fig. 10.3 and the circles show the results for a uniformly bright star $(u_\lambda = 0)$ of diameter θ_{UD}. The ratio θ_{UD}/θ_{LD} was chosen so that the shapes of the two curves can be compared. It can be seen that they are very similar; the only significant differences appear in the secondary lobes and do not exceed about $0\cdot01$ of the zero-baseline correlation. It follows that these two extreme cases of zero and complete limb-darkening cannot be distinguished unless the signal/noise ratio is high (~ 100 to 1 at zero baseline). At Narrabri this performance was only reached on the two brightest stars (α C Ma and α Car) and so, for most stars, it is reasonable to say that the interferometer measured only

123

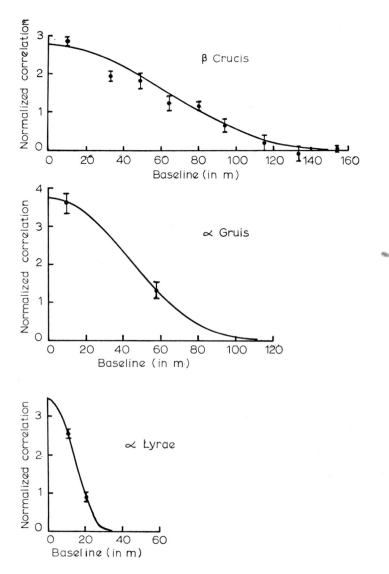

Fig. 10.2.   Correlation versus baseline measured for three stars of different angular size.   The full lines were fitted to the points as described in the text.   From Hanbury Brown, Davis, Allen and Rome (1967).

124

the angular diameter of the equivalent uniform disc—the uniform disc being defined as the disc which radiates the same total light flux as the actual star and which has a Fourier transform for the equivalent line source which closely matches that of the actual star.

However, it should be noted that the angular diameter of the equivalent uniform disc does not differ much from the true angular

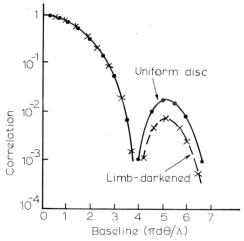

Fig. 10.3. The effect of limb-darkening on the variation of correlation with baseline. $\times$: complete limb-darkening $u_\lambda=1$, $\bullet$: uniform disc $u_\lambda=0$. From Hanbury Brown and Twiss (1958 a).

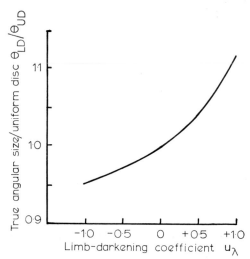

Fig. 10.4. The effect of limb-darkening on the apparent angular diameter of a star. $\theta_{\mathrm{LD}}/\theta_{\mathrm{UD}}$ is the ratio of the true angular size to the equivalent uniform disc, u_λ is the limb-darkening coefficient. From Hanbury Brown and Twiss (1958 a).

125

diameter of a star even in the case of complete limb-darkening. Thus Hanbury Brown and Twiss give the ratio θ_{LD}/θ_{UD} (true angular size/uniform disc) as

$$\theta_{LD}/\theta_{UD} = [(1 - u_\lambda/3)/(1 - 7u_\lambda/15)]^{1/2} \tag{10.13}$$

for the limb-darkening law of equation (10.12). This ratio is shown in fig. 10.4 as a function of u_λ.

To summarize, the intensity interferometer yielded the angular diameter (θ_{UD}) of the equivalent uniform disc of a single star. For most stars the true angular size of the limb-darkened disc (θ_{LD}) cannot be found from the observations but must be derived by introducing a small theoretical correction which depends on the assumed law of limb-darkening.

10.3.3 The effects of multiple stars

The procedure for finding the angular diameter of a star as described above is based on the assumption that we are dealing with a single and not a binary or multiple star. It is of course possible to reject known binaries on the basis of the available optical evidence. However, experience shows that this is not good enough and even some of the well-known bright stars have proved to be multiple when observed with the interferometer. It is therefore important, before a theoretical curve is fitted to the measurements, to check whether or not they are consistent with a single star.

Let us first consider the simple case when the interferometer observes a binary star in which the components have an angular separation θ_s, which is too small to be resolved by the individual reflectors. If the line joining these two components makes an angle ψ with the baseline, the shape of the curve relating correlation to baseline is given by

$$\Gamma^2(d) = \frac{1}{(I_1{}^2 + I_2{}^2)} [I_1{}^2 \Gamma_1{}^2(d) + I_2{}^2 \Gamma_2{}^2(d)$$
$$+ 2I_1 I_2 |\Gamma_1(d)||\Gamma_2(d)| \cos \{2\pi\theta_s d \cos \psi/\lambda\}] \tag{10.14}$$

where I_1 and I_2 are the intensities of the two components, both of which are assumed to be too small to be resolved by the individual reflectors. This curve is modulated at a frequency determined by $\theta_s \cos \psi$ and the observed correlation oscillates between values which are proportional to

$$[I_1|\Gamma_1(d)| + I_2|\Gamma_2(d)|]^2 \text{ and } [I_1|\Gamma_1(d)| - I_2|\Gamma_2(d)|]^2. \tag{10.15}$$

Since $\theta_s \cos \psi$ varies with time as the position angle changes, the correlation observed at a given baseline also varies with time and has a mean value which lies between these extremes; it is therefore less

126

than the correlation expected from a single star with the same light flux. In the more complicated case where the star has more than two components the correlation will also vary with time but will be further reduced relative to a single star giving the same total light flux.

In principle, it is therefore possible to distinguish a multiple from a single star by observing that the correlation is less than expected from a single star of the same brightness, or by noting that the normalized correlation varies with time, or with baseline, in a way which is inconsistent with a single star. As an example, consider the case where the separation of a binary star θ_s is such that it is completely resolved at the shortest baseline used $(d_{min} \gg \lambda/2\theta_s)$. Then the normalized correlation, averaged over a wide range of position angles, is given by

$$\Gamma^2(d) = \frac{1}{(I_1^2 + I_2^2)} [I_1^2 \Gamma_1^2(d) + I_2^2 \Gamma_2^2(d)]. \qquad (10.16)$$

Thus, in this simple case the interferometer treats the two stars as separate entities and the observed correlation is the sum of the correlation due to each star separately. It follows that, at short baselines where neither of the components is individually resolved, the correlation will be reduced, relative to a single star, by the factor

$$(I_1^2 + I_2^2)/(I_1 + I_2)^2. \qquad (10.17)$$

Similarly, if the star had n components and the separations between them were all resolved, the correlation would be reduced, relative to a single star, by the factor

$$\sum_n I^2/(\sum_n I)^2. \qquad (10.18)$$

In these cases we can distinguish a multiple from a single star by measuring $\overline{c_N(0)}$ the normalized zero-baseline correlation, and comparing it with the value for a single star.

A more complicated case is where the separations θ_s between the components of a multiple star are not resolved at the shortest baseline. In this case it may happen that the correlation at short baselines is not reduced relative to a single star, and the only way of detecting that such a star is not single would be to make observations at several baselines and to compare the curve with that expected from a single star. However, this method is very time-consuming and is only practicable for a few very bright stars; we therefore used the following procedure.

When seeking to measure single stars, selected on the best available optical evidence, the first measurements were always made at the shortest possible baseline (10 m). If the observed correlation was less than expected and would obviously lead to a lower value of $\overline{c_N(0)}$ than that given by a single star, the star was deleted from the programme of single stars.   The practical criterion adopted was that if the star gave less than 80 per cent of the correlation expected from a single star, it was rejected; this implied that, if the star is binary, then the secondary component is at least 2·2 magnitudes fainter than the primary.   As a further safeguard the optical and spectroscopic data were reviewed and, if there was a suspicion that the star might be multiple, observations were made at three or more baselines and the ratios of these correlations were compared with those expected from a single star.

Experience showed that this is an effective test of whether a star is single or not—a single star being defined as one in which any secondary component is at least 2·2 magnitudes fainter.   For example, it was immediately obvious that well-known spectroscopic binaries such as λ Sco and α Vir are not single; furthermore, the interferometer showed beyond question that some of the well-known bright 'single' stars are multiple (e.g. σ Sgr, δ Sco).   Nevertheless, unless it was convenient to spend a long time on each star, there was inevitably a small residual uncertainty as to whether a star is single or not.

10.3.4   *The effects of stellar rotation*

One of the many factors which was considered when fitting theoretical curves to observations of single stars is rotation.   The most obvious effect of rotation is to change the shape of a star, increasing the equatorial diameter relative to the polar diameter.   This means that we are no longer dealing with a circular disc and, if the surface brightness remained uniform, these changes would appear directly as apparent changes of the angular diameter which would, in general, vary with position angle.   However, the situation is complicated by the fact that the variation of surface gravity over a rotating star leads, at least in theory, to significant variations of surface brightness.   For example, the reduction of surface gravity at the Equator may be expected to lead to comparatively low brightness in that region.   The overall effect of rotation on the apparent angular size will therefore depend in a complicated way on the orientation of the rotation axis relative to the baseline of the interferometer.

The effects of rotation have been discussed briefly by Hanbury Brown, Davis, Allen and Rome (1967) and at greater length by Johnston and Wareing (1970).   In the first of these papers the authors consider a very rapidly rotating star with an equatorial velocity of 350 km s^{-1} observed in three different orientations.   The model of this star is based on that given by Ireland (1966); it assumes simple solid rotation in hydrostatic equilibrium, the radiated flux at any point on the surface being given by von Zeipel's theorem as proportional to the local value of gravity.   When the rotation axis of the star is parallel to the baseline, numerical integration of the brightness distribution shows that the equivalent strip source is both narrower and limb-brightened as compared with a uniform disc with a radius equal to the

128

mean of the polar and equatorial radii. But the Fourier transform of the distribution is almost identical to that of this uniform disc, at least out to the first minimum in the transform, and it follows that the measured angular diameter is almost unchanged by rotation. This rather surprising result is due to the fact that changes in the Fourier transform due to the reduction in polar diameter are, to a first order, cancelled by those produced by polar brightening. When the star is viewed equator-on, with the rotational axis perpendicular to the base-line, the calculations show that the angular diameter is apparently 8 per cent greater than the uniform disc; when it is viewed pole-on the angular diameter appears to be about 5 per cent greater than the uniform disc.

Johnson and Wareing consider the two rapidly rotating stars Regulus (α Leo) and Altair (α Aql). Using simple models, based again on von Zeipel's theorem, they find that the maximum apparent change of angular diameter with aspect is roughly 6 per cent for Regulus and 4 per cent for Altair.

Both these discussions lead to the same conclusion that, for most stars, the effects of rotation are so small that they would have been difficult to detect with the Narrabri interferometer. They were not therefore likely to be a significant source of error in the Narrabri programme. However, for very rapidly rotating stars, viewed in a favourable aspect and over a sufficient range of position angles, it should be possible to detect the effects of rotation and this is confirmed by the measurements on Altair reported in § 11.7.

Although the effects of rotation are not likely to be significant in the work of the interferometer, where angular diameters were measured with a precision of about ± 5 per cent, it is clear that they would have to be taken into account in more precise work with a more sensitive instrument. Furthermore, it is interesting to note that these effects may be comparable with those of limb-darkening (§ 10.3.2).

10.3.5 *The effects of polarization*

It is interesting to enquire if the angular size of a single star depends upon whether it is observed in polarized or unpolarized light. If scattering by electrons plays a significant role in the star's atmosphere, then it is to be expected that the limb-darkening law will depend upon the plane of polarization.

Chandrasekhar (1946) has analysed pure scattering by free electrons in a semi-infinite plane parallel atmosphere, and has shown that the radiation is polarized and that the percentage polarization increases towards the limb where it reaches 11 per cent. The ratios of the light flux at the centre of the star to that at the limb are 0·36 and 0·29 for the two planes of polarization. If therefore we measure the star in light polarized parallel $E\|$ and perpendicular $E\bot$ to the baseline of the

K

interferometer, we shall find the angular size to be larger for $E\perp$ than for $E\|$.

A later paper by Harrington (1970) discusses the same problem in greater detail and gives the variation of polarization over the stellar disc as a function of the wavelength and the ratio of scattering to absorption. His results indicate that, for a wavelength of 443·0 nm, for the range of spectral types (O to F8) observed at Narrabri and for a finite amount of absorption, we may take the polarization given by Chandrasekhar as an upper limit.

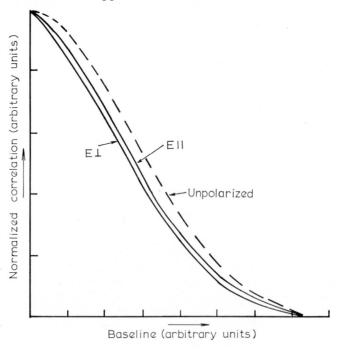

Fig. 10.5. The effect of electron-scattering in the corona of a star on observations made in polarized light. $E\perp$, $E\|$ correspond to polarization perpendicular and parallel to the baseline of the interferometer. The broken line shows the shape of the curve for unpolarized light.

A rough estimate of the effect of this polarization can be made by treating it as a change in the limb-darkening coefficient u_λ. Thus, if we represent the difference between the distributions for the two planes of polarization as a change in u_λ of $(0·39-0·26)=0·07$, then fig. 10.4 shows that the corresponding change in the apparent angular diameter of the star would be less than 1 per cent. Since this estimate is an upper limit we may conclude that any effects of this polarization are unlikely to have been significant in the measurements of single stars made at Narrabri.

130

Another possibility which we have considered is that electron scattering in an ionized corona, possibly associated with mass loss from a hot star, might produce observable effects. A thorough theoretical analysis of this question has not yet been made. We have considered only one elementary model in which the corona of a star is formed by fully ionized hydrogen streaming radially away at a constant velocity. Light emitted from the star is scattered by electrons in this corona and is polarized: it is assumed that the optical depth is so small that photons leaving the star never suffer more than one collision on their path through the corona. Fig. 10.5 shows curves of correlation versus baseline calculated for this model for two planes of polarization; as one would expect the angular diameter is greater with the polarization normal to the baseline; the curves were calculated for $x = 0.2$ where x is a scattering parameter given by

$$x = \frac{S\sigma_0}{4\pi m_H VR} \approx \frac{0.3S}{VR} \tag{10.19}$$

where S is the rate of mass loss in units of $10^{-6} M_\odot \, \mathrm{yr}^{-1}$ ($M_\odot = \mathrm{mass}$ of sun), σ_0 is the Thomson scattering cross-section of the electron, V is the velocity of mass efflux in units of $10^3 \, \mathrm{km \, s}^{-1}$ and R is the radius of the star expressed in solar radii.

From observations of three hot super-giants (δ Ori, ϵ Ori, ζ Ori), Morton (1967) suggests that they lose mass at about $10^{-6} M_\odot \mathrm{yr}^{-1}$ with an efflux velocity of about 1400 km s^{-1}. Substituting these figures in equation (10.19) and putting $R = 50$ solar radii, the scattering parameter $x = 0.004$ and it can be shown that the corresponding change in apparent angular size with polarization is about 0·1 per cent. It therefore seems unlikely that effects of mass loss from hot stars will be observable.

Although this discussion suggests that any change in the apparent angular diameter of a star with plane of polarization is likely to be less than 1 per cent, we decided to observe the bright super-giant Rigel in polarized light. The results are presented in §11.8.

CHAPTER 11

results

11.1 *Angular Diameters and Zero-baseline Correlations*

The main programme of the interferometer at Narrabri was the measurement of 32 stars. A list of these stars is shown in Table 11.1. They were chosen to be as widely representative of different types of star as possible and their distribution in spectral type and luminosity class is shown in Fig. 12.1. The stars are all brighter than about $B = +2 \cdot 5$ and are limited to declinations south of about 20° N in order to avoid excessive atmospheric extinction at low elevations, the only exception being the bright star Vega (α Lyr) at 39° N. Their spectral range was limited at one end to stars hotter than type F8 by considerations of the signal/noise ratio, which is a function of surface temperature as discussed in § 5.6.

The final results are given in Table 11.1. Columns 1–3 identify the star together with its spectral type and luminosity class. Columns 4 and 5 give the results of fitting theoretical curves to the normalized correlations observed at different baselines following the procedure described in § 10.3. Column 4 shows the normalized zero-baseline correlation C_N corrected for the effects of partial resolution ($\Delta\lambda = 1$) and expressed as the ratio of the measured correlation to that expected (see § 11.3) from a single star; σ is the associated r.m.s. uncertainty estimated by the method discussed in § 10.2. Column 5 shows the apparent angular diameter θ_{UD} of the *equivalent uniform disc* as discussed in § 10.3.1 with its r.m.s. uncertainty σ.

One of the problems in selecting these stars was to avoid unsuspected double stars which might give misleading results. From the existing optical data one cannot be sure whether or not any particular star is single and before including it in the programme it was necessary to measure the zero–baseline correlation; in this way it is possible (see § 10.3.3) to assign an upper limit to the brightness of any companion star. For the majority of the stars in Table 11.1 the zero-baseline correlation does not differ significantly from unity and, taking the r.m.s. uncertainties in C_N to be about $\pm 0 \cdot 1$, one can say that any companion stars are likely to be at least 2·5 magnitudes fainter than their primaries. It follows that these companions, if they exist, do not contribute more than 1 per cent of the observed short-baseline correlations and it is therefore unlikely that they cause any significant errors in the measured angular diameters of the primary stars.

132

However, there are five stars in the list (α Vir, γ^2 Vel, ζ Ori, β Cru, δ Sco) which are definitely multiple. Two of these stars (α Vir, γ^2 Vel) are well-known binaries. For both these stars the variation of correlation with time, position angle and baseline was compared in a computer with a theoretical model of a binary star (§ 11.4, § 11.5) and the value of the angular size of the primary star was found. The resulting values of θ_{UD} are shown in Table 11.1 but, since the separation between the components was not fully resolved at short baselines, no values of c_N are quoted. Three of the stars in Table 11.1 (ζ Ori, β Cru, δ Sco) have values of c_N which are significantly less than unity and this implies that they are multiple. However, in each case it can be shown that any error in the angular size of the primary due to a companion star is likely to be considerably less than the uncertainty in the measurement shown in the table. We may therefore conclude that the angular diameters in Table 11.1 all refer to single stars; in some cases, notably the five multiple stars mentioned above, the angular size refers to the brightest star in the system.

11.2 Radii, Fluxes and Temperatures

11.2.1 True angular diameters

The observational data presented in column 5 of Table 11.1 give the angular diameter of each star in terms of θ_{UD} the angular diameter of the equivalent uniform disc. The true angular diameter θ_{LD} is greater than this value due to the effects of limb-darkening as outlined in § 10.3.2. For the conventional cosine law of limb-darkening (equation (10.12) the ratio θ_{LD}/θ_{UD} is given by equation (10.13) which can be used to evaluate θ_{LD} for all the stars. However, the values of θ_{LD} shown in column 6 were derived by a more precise method in which the limb-darkening was taken from model atmospheres of the appropriate temperature and gravity following Hanbury Brown, Davis and Allen (1974).

11.2.2 Radii

We have used these values of θ_{LD} to find a basic parameter, the physical radius R, of 15 of the stars in Table 11.1. for which the distance is reasonably well known. The resulting values of R are shown in Table 11.2; they are expressed in units of ($R_\odot$) the solar radius (Allen, 1963).

For 12 of these stars the radius was found from θ_{LD} and the trigonometrical parallax (Jenkins, 1963); 11 of the stars were chosen because the probable error in their parallaxes is quoted as less than ± 25 per cent; 1 star (α Car) was included, although the quoted uncertainty in its parallax is ± 35 per cent, because of the paucity of information about supergiants. For 3 stars (α Vir, γ^2 Vel, ζ Pup) the parallax was derived from other sources; α Vir is discussed in § 11.4 and γ^2 Vel in

133

1	2	3	4	5	6	7
			Zero-baseline correlation	Angular diameter $\times 10^{-3}$ sec of arc		Temperature
Star number	Star name	Type	$c_N \pm \sigma$	$\theta_{UD} \pm \sigma$	$\theta_{LD} \pm \sigma$	$[T_e(F) \pm \sigma]/K$
472	α Eri	B 3 (Vp)	0·98±0·05	1·85±0·07	1·92±0·07	13 700±600
1713	β Ori	B 8 (Ia)	0·98±0·08	2·43±0·05	2·55±0·05	11 500±700
1790	γ Ori	B 2 (III)	1·03±0·07	0·70±0·04	0·72±0·04	20 800±1300
1903	ε Ori	B O (Ia)	0·86±0·07	0·67±0·04	0·69±0·04	24 500±2000
1948	ζ Ori	O 9·5 (Ib)	0·60±0·06	0·47±0·04	0·48±0·04	26 100±2200
2004	κ Ori	B 0·5 (Ia)	1·18±0·09	0·44±0·03	0·45±0·03	30 400±2000
2294	β CMa	B 1 (II–III)	1·07±0·08	0·50±0·03	0·52±0·03	25 300±1500
2326	α Car	F 0 (Ib–II)	0·75±0·22	6·1±0·7	6·6±0·8	7500±250
2421	γ Gem	A 0 (IV)	1·17±0·09	1·32±0·09	1·39±0·09	9600±500
2491	α CMa	A 1 (V)	0·91±0·06	5·60±0·15	5·89±0·16	10 250±150
2618	ε CMa	B 2 (II)	0·89±0·06	0·77±0·05	0·80±0·05	20 800±1300
2693	δ CMa	F 8 (Ia)	0·93±0·18	3·29±0·46	3·60±0·50	—
2827	η CMa	B 5 (Ia)	0·99±0·09	0·72±0·06	0·75±0·06	14 200±1300
2943	α CMi	F 5 (IV–V)	0·98±0·10	5·10±0·16	5·50±0·17	6500±200
3165	ζ Pup	O 5 (f)	1·04±0·08	0·41±0·03	0·42±0·03	30 700±2500
3207	γ² Vel	WC 8+O 9 (I)	—	0·43±0·05	0·44±0·05	29 000±3000
3685	β Car	A 1 (IV)	1·01±0·06	1·51±0·07	1·59±0·07	9500±350
3982	α Leo	B 7 (V)	1·12±0·07	1·32±0·06	1·37±0·06	12 700±800
4534	β Leo	A 3 (V)	1·17±0·10	1·25±0·09	1·33±0·10	9050±450

Bright star catalogue number	Star name	Spectral type and luminosity class	Zero-baseline correlation (Δλ=1)	Angular diameter of equivalent uniform disc	True angular diameter	Theoretical effective temperature
4662	γ Crv	B 8 (III)	0·97 ± 0·10	0·72 ± 0·06	0·75 ± 0·06	13 100 ± 1200
4853	β Cru	B 0·5 (III)	0·88 ± 0·03	0·702 ± 0·022	0·722 ± 0·023	27 900 ± 1200
5056	α Vir	B 1 (IV)	—	0·85 ± 0·04	0·87 ± 0·04	22 400 ± 1000
5132	ε Cen	B 1 (III)	1·02 ± 0·07	0·47 ± 0·03	0·48 ± 0·03	26 000 ± 1800
5953	δ Sco	B 0·5 (IV)	0·75 ± 0·07	0·45 ± 0·04	0·46 ± 0·04	—
6175	ζ Oph	O 9·5 (V)	1·01 ± 0·12	0·50 ± 0·05	0·51 ± 0·05	—
6556	α Oph	A 5 (III)	0·94 ± 0·09	1·53 ± 0·12	1·63 ± 0·13	8150 ± 400
6879	ε Sgr	A 0 (V)	1·02 ± 0·06	1·37 ± 0·06	1·44 ± 0·06	9650 ± 400
7001	α Lyr	A 0 (V)	0·99 ± 0·04	3·08 ± 0·07	3·24 ± 0·07	9250 ± 350
7557	α Aql	A 7 (IV, V)	0·94 ± 0·06	2·78 ± 0·13	2·98 ± 0·14	8250 ± 250
7790	α Pav	B 2·5 (V)	1·01 ± 0·07	0·77 ± 0·05	0·80 ± 0·05	17 100 ± 1400
8425	α Gru	B 7 (IV)	1·11 ± 0·08	0·98 ± 0·07	1·02 ± 0·07	14 800 ± 1200
8728	α PsA	A 3 (V)	1·02 ± 0·08	1·98 ± 0·13	2·10 ± 0·14	9200 ± 500

1. Bright star catalogue number (Hoffleit, 1964).
2. Star name.
3. Spectral type and luminosity class (in brackets).
4. Zero-baseline correlation normalized by the value expected from a single unresolved star and corrected for partial resolution (Δλ=1) (§ 10.3.1).
5. Angular diameter of equivalent uniform disc with r.m.s. uncertainty (§ 10.3.1).
6. True angular diameter allowing for the effects of limb-darkening (§10.3.2 and § 11.2.1).
7. Theoretical effective temperature of star computed by Webb (1971).

Table 11.1.   The angular diameters and zero-baseline correlations of 32 stars.

135

Star	Spectral type	Luminosity class	Radius $(R_\odot^{-1})$ $R \pm \sigma_R$
γ^2 Vel	WC 7+O 7	—	$16\cdot3 \pm 2\cdot9$
ζ Pup	O 5f	—	$15\cdot6 \pm 2\cdot2$
α Vir	B 1	V	$7\cdot9 \pm 0\cdot7$
α Gru	B 5	V	$2\cdot2 \pm 0\cdot6$
α Leo	B 7	V	$3\cdot8 \pm 1\cdot0$
α Lyr	A 0	V	$2\cdot8 \pm 0\cdot2$
γ Gem	A 0	IV	$4\cdot2 \pm 0\cdot7$
α CMa	A 1	V	$1\cdot69 \pm 0\cdot05$
β Car	A 1	IV	$4\cdot5 \pm 1\cdot8$
β Leo	A 3	V	$1\cdot9 \pm 0\cdot2$
α PsA	A 3	V	$1\cdot6 \pm 0\cdot2$
α Oph	A 5	III	$3\cdot1 \pm 0\cdot5$
α Aql	A 7	IV, V	$1\cdot65 \pm 0\cdot09$
α Car	F 0	I b–II	42 ± 22
α CMi	F 5	IV–V	$2\cdot1 \pm 0\cdot1$

NOTE
Luminosity class

I a, b=super-giants
II=bright giants
III=giants
IV=sub-giants
V=main sequence, dwarfs

Table 11.2.   The radii of 15 stars.

§ 11.5; the parallax of ζ Pup has been taken as equal to that of γ^2 Vel following Davis, Morton, Allen and Hanbury Brown (1970).

11.2.3 *Emergent fluxes*

Another important parameter of a star is the absolute monochromatic flux (F_λ) emitted by the stellar surface or, in other words, the emergent flux.   For the stars in Table 11.1 this was found from the relation

$$F_\lambda = 4f_\lambda/\theta_{LD}{}^2 \qquad (11.1)$$

where f_λ is the measured absolute monochromatic flux received from the star at a wavelength λ.

These values of F_λ are fundamental to the comparison of actual stars with the theory of stellar atmospheres.   It is worth noting that, apart from a minor correction for limb-darkening, they are independent of theoretical models and based simply on observational data.   They were used in the next section to find effective temperatures.

136

11.2.4 *Effective temperatures*

The effective temperature of a star is defined in terms of the emergent flux F_λ by the relation

$$\int_0^\infty F_\lambda \, d\lambda = \sigma T_e^4 \qquad (11.2)$$

where σ is Stefan's constant.   If therefore we know F_λ as a function of λ over the complete spectrum of the star we can find a value of T_e which is based entirely on observational data; we shall call this the *empirical effective temperature* $T_e(\text{emp})$.   This can, of course, only be done for stars which are sufficiently cool so that their spectral distribution can be measured at the surface of the Earth without significant errors due to loss of ultra-violet radiation in the upper atmosphere.   It is also possible for the comparatively few hot stars for which measurements of ultra-violet flux have been made from rockets or satellites.

At the present date there are sufficient data to establish empirical effective temperatures for only five stars measured by the interferometer. The results are shown in column 4 of Table 11.3.   These temperatures are of particular value because they are based entirely on observational data and not on theoretical models; it is to be expected that, as more rocket and satellite measurements become available, it will be possible to extend the list.

For the majority of stars in Table 11.1 the complete spectral distribution is not yet available and the effective temperatures have been found by matching the values of F_λ (§ 11.2.3) to a grid of theoretical models calculated for a range of effective temperatures; we shall call these *theoretical effective temperatures* T_e *(F)*.   These temperatures

1 Star	2 Spectral type	3 Luminosity	4 Empirical $[T_e(\text{emp}) \pm \sigma_{\text{r.m.s.}}]/\text{K}$	5 Theoretical $[T_e(F) \pm \sigma_{\text{r.m.s.}}]/\text{K}$
α CMi	F 5	IV–V	6475 ± 240	6500 ± 200
α Car	F 0	I b–II	7425 ± 250	7500 ± 250
α Aql	A 7	IV–V	8120 ± 250	8250 ± 250
α CMa	A 1	V	9680 ± 230	10250 ± 150
β Ori	B 8	I a	$.11250 \pm 460$	11500 ± 700

1. Star name
2. Spectral type
3. Luminosity class
4. Effective empirical temperature $T(\text{emp})/\text{K}$ (§ 11.2.4)
5. Theoretical effective temperature $T(F)/\text{K}$ (§ 11.2.4)

Table 11.3.   Comparison of empirical and theoretical effective temperatures.
From Webb (1971).

have been derived for all the stars in Table 11.1. Since each theoretical model is characterized by two parameters, temperature and gravity, a single value of F_λ is not sufficient by itself to select a model. Three additional observed parameters were used as auxiliary information, C, the mean slope of the Paschen continuum around 550 nm, D, the size of the Balmer jump and W, the equivalent width of Hγ. Taking the four parameters in turn (F, C, D, W), the locus of temperature and gravity was plotted for all models showing the observed value of that parameter. Examples are shown in fig. 11.1 where the loci are plotted in terms of $\log g$ and the reciprocal effective temperature θ_e. Wherever possible these plots were used to choose a surface gravity for each star which gave the best agreement between the most sensitive parameters. The final theoretical effective temperature was then read directly from the F locus using the chosen gravity. It can be seen from the curves that F is the most gravity-independent and temperature-sensitive of the parameters. In a few cases the gravity could not be found in this way and was estimated from the M.K. luminosity class. The final values of theoretical effective temperature ($T_e(F)$) are shown in column 7 of Table 11.1.

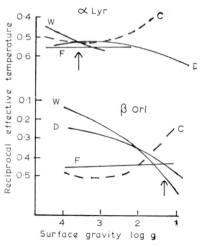

Fig. 11.1.   Effective temperature versus surface gravity for model atmospheres which give the observed values of F (emergent flux), C (slope of Paschen continuum), D (Balmer jump), and W (equivalent width of H$_y$) for two stars. The temperature scale is in conventional units of reciprocal temperature ($5040/T_e$ K^{-1}). From R. J. Webb (1971).

It is interesting to enquire how well these theoretical temperatures, which rely heavily on models, agree with the entirely empirical temperatures shown in Table 11.3. For convenience the theoretical temperature for each star has been shown in column 5 of Table 11.3 and it can be seen that the agreement is good. It is also interesting to

138

enquire how closely the scale of temperatures derived from the inter-ferometer agrees with other scales. The temperatures $T_e(F)$ in Table 11.1 given by the interferometer are shown in fig. 11.2, together with the well-known scale of Harris (1963). It can be seen that there is general agreement but there are some differences. A discussion of the significance of these differences and of the relation of the temperature scale to the theoretical models is beyond the scope of this book.

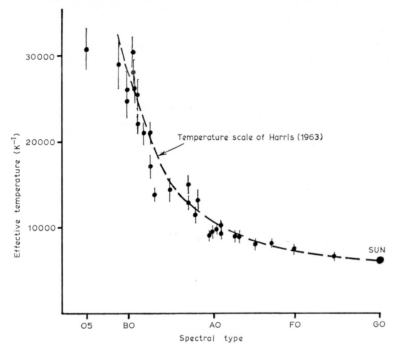

Fig. 11.2. Comparison of scale of effective temperature. The points show the results obtained with the interferometer at Narrabri with their r.m.s. uncertain-ties (from Webb, 1971); the broken line shows the scale proposed by Harris (1963).

11.3   *The Detection of Multiple Stars*

In § 10.3.3 we reviewed the effects of multiple stars on the observed correlation. We showed that it is possible to distinguish a multiple from a single star by observing that the correlation is less than for a single star of the same brightness or that it varies with time or baseline in a way which is inconsistent with a single star.

The main problem in putting this scheme into practice was to establish the correlation expected from a single star. One major source of uncertainty were the modifications to the equipment which we introduced over a period of years (mainly new phototubes). These

139

modifications inevitably changed the zero-baseline correlation from a single star and so after each modification we had to recalibrate the equipment by observing a standard star; as transfer standards we used α Eri and β Cru. A second and more problematic source of uncertainty was that we could not be sure that any particular star was single; for example, one of our transfer standards β Cru eventually proved to be multiple.

We finally established the zero-base line correlation for a single star by analysing the distribution of the measured values of c_N for all stars believed to be single. This distribution showed that three of the stars in the list (ζ Ori, β Cru, δ Sco) have significantly low values of c_N and are therefore multiple. When these stars were removed the remaining values of c_N proved to be distributed about their mean with a dispersion in satisfactory agreement with the uncertainties in their individual values. This mean was therefore taken to be the zero-baseline correlation expected from a single unresolved star and was used in finding the values of C_N shown in Tables 11.1 and 11.4. The uncertainty in this mean is about ± 1.5 per cent.

B.S.	Name	$C_N \pm \sigma^*$	Remarks
1948/9	ζ Ori	0.60 ± 0.06	
3207	γ^2 Vel	—	See § 11.5
3485	δ Vel	0.65 ± 0.06	
4853	β Cru	0.88 ± 0.03	
5056	α Vir	—	See § 11.4
5267	β Cen	0.47 ± 0.02	
5953	δ Sco	0.75 ± 0.07	
6527	λ Sco	0.48 ± 0.08	
7121	σ Sgr	0.54 ± 0.07	

* Zero-baseline correlation with effects of partial resolution removed ($\Delta_\lambda = 1$) normalized by value expected from a single unresolved star.

Table 11.4. Multiple stars observed with the interferometer.

Table 11.4 is a list of nine multiple stars which were observed. Five of them (ζ Ori, γ^2 Vel, β Cru, α Vir, δ Sco) are in Table 11.1. ζ Ori is listed (Hoffleit, 1964) as a triple system with a difference of about 2 magnitudes between the brightest components. The observed value of C_N (0.60 ± 0.06) shows that the brightest component must itself be a double star and the simplest interpretation is that it is a binary star with $\Delta m \simeq 2$ mag. It has previously been suspected by some observers that β Cru may be a double star; the value of C_N (0.88 ± 0.03) shows definitely that it is multiple and is consistent with a binary star with $\Delta m \simeq 2.9$ mag. δ Sco has not been listed previously as a multiple star but the observed value of C_N (0.75 ± 0.07) is consistent with a binary star with $\Delta m \simeq 1.9$ mag. The two well-known binary stars (α Vir, γ^2 Vel) were observed for special reasons and the work is described in §11.4 and §11.5.

Four of the stars in Table 11.4 (δ Vel, β Cen, λ Sco, σ Sgr) do not appear among the 32 stars in Table 11.1. It is interesting to note that all these well-known stars were possible candidates for our observing list but were rejected because the interferometer showed them to have companions too bright for our programme on single stars. δ Vel is listed (Hoffleit, 1964) as having a faint double companion, however, the observed value of C_N (0.65 ± 0.06) shows that the bright component must itself be multiple; the simplest interpretation is that it is a binary with $\Delta m \simeq 1.3$ mag. β Cen is well-known to exhibit a variable radial velocity and it is interesting to note that the value of C_N (0.47 ± 0.02) is consistent with a binary star with two equally bright components. λ Sco was reported many years ago to be a spectroscopic binary but the brightness of the companion was not known; our measurement of C_N (0.48 ± 0.08) confirms that it is multiple and suggests a binary star with components of equal brightness. σ Sgr is a well-known bright star which was thought to be single, but the observed value of C_N (0.54 ± 0.07) is consistent with a binary star with components of roughly equal brightness ($\Delta m \simeq 0.6$ mag).

It is interesting to note how many of the well-known bright stars in our original observing list were found to be multiple. The result suggests that more extensive observations of the zero-baseline correlation for stars would contribute significantly to the population statistics of multiple stars. In principle it would have been possible to observe all these multiple stars in greater detail and find out more about them, for example, whether they are double or triple, but without a more sensitive interferometer such a programme would have taken far too long.

11.4   The Spectroscopic Binary: Spica (α Vir)

11.4.1   Introduction

In principle, observations of a binary star by an interferometer, if made at suitable baselines and times, can yield the angular diameter and brightness ratio of the components and their angular separation as a function of time. From this information alone it is possible, again in principle, to find the angular size of the semi-major axis of the relative orbit, the eccentricity, the time of periastron passage, the longitude of the line of apsides, the inclination of the orbit, the position angle of the line of nodes, the period of the orbit and the sense of orbital motion. If these data are then combined with conventional photometric and spectroscopic observations it is possible, still in principle, to find the *distance*, absolute magnitude, mass, radius, surface gravity and temperature of both stars.

141

It is therefore clear that observations of binary stars with an interferometer are, potentially, of considerable importance to astronomy. Any technique which promises to establish reasonably precise distances beyond the limits of trigonometrical parallax and to extend our meagre knowledge of the masses, radii and absolute magnitudes of stars, ought to be explored. Unfortunately the interferometer at Narrabri was not sufficiently sensitive to measure a reasonable number of binary stars; however, the well-known non-eclipsing double-lined spectroscopic binary star Spica (α Vir) is sufficiently bright and provided an opportunity to test the technique.

Two series of observations, made in 1966 and 1970, have been described by Herbison-Evans, Hanbury Brown, Davis and Allen (1971). A summary of their work follows.

11.4.2 *Method of observation*

The observations were carried out using the standard observing procedure already described in § 9.5. Briefly, the two reflectors were guided to follow α Vir and their separation—the baseline—was maintained constant and perpendicular to the direction of the star. The correlation was measured as usual, over intervals of 100 s, together with the light fluxes received by the two reflectors. The observed correlation was normalized in the usual way (§ 10.1) to find the *normalized correlation* $\overline{c_N(d)}$ for each of these 100 s intervals. Typically the observations at any one baseline lasted for several hours.

11.4.3 *Analysis of observations*

A computer programme was written to calculate the theoretical correlation as a function of time expected from a binary star with a given set of assumed parameters. These calculated correlations were then compared with the observations by the computer and the assumed parameters were varied in an iterative process to obtain the best possible fit.

The programme assumed that the component stars present uniformly bright circular discs. Under these conditions, following equation (10.12), the normalized correlation is given by

$$\overline{c_N(d)} = \frac{C}{(1+\beta)^2} [\beta^2 \Gamma_1^2(d) + \Gamma_2^2(d)$$
$$+ 2\beta |\Gamma_1(d)||\Gamma_2(d)| \cos(2\pi\theta_s d \cos\psi/\lambda_0)] \tag{11.3}$$

where

$$\Gamma_1(d) = 2J_1(\pi\theta_{UD1}d/\lambda_0)/(\pi\theta_{UD1}d/\lambda_0) \tag{11.4}$$

and

$$\Gamma_2(d) = 2J_1(\pi\theta_{UD2}d/\lambda_0)/(\pi\theta_{UD2}d/\lambda_0) \tag{11.5}$$

β is the brightness ratio of the components and λ_0 is the wavelength; $\theta_{UD1}, \theta_{UD2}$ are the angular diameters (equivalent uniform discs) of the primary and secondary; θ_s is the angular separation of the components projected onto the plane of the sky; ψ is the angle in the plane of the sky between the projection of the line joining the stars and the baseline of the interferometer; C is an instrumental constant corresponding to the normalized correlation to be expected from an unresolved single star. The parameters of the orbit which enter the calculation are θ_a the angular semi-major axis; i the inclination of the orbital plane; Ω the position angle of the line of nodes; e the eccentricity; T the epoch of periastron passage; ω the longitude of the line of apsides; P the period of the orbit measured from periastron to periastron; U the period of rotation of the line of apsides. The senses of rotation of the stars in their orbit and of the line of apsides also enter the calculation.

As a first step in the analysis T, e, P, ω, U and θ_{UD2} were fixed. The values of T, e, P, and ω were taken from spectroscopic observations and are shown in Table 11.6; although they can also be found from the interferometer in the present case, the spectroscopic values are more precise. The value of θ_{UD2}, the angular size of the fainter component, was estimated because the signal/noise ratio of the interferometer was not high enough to yield it with acceptable accuracy from the observations. To simplify the calculations, with no significant loss of precision, it was assumed that the line of apsides did not rotate over the comparatively short period of each set of observations ($U = \infty$) and the value of ω, the position angle of this line, was given fixed values appropriate to the mean epochs of the observations in 1966 and 1970.

The remaining six unknown 'free' parameters (i, θ_a, β, Ω, θ_{UD1}, and C) were then found by the computer as follows. The difference between the observed and computed correlations was found for each interval, squared and weighted by the square of its signal/noise ratio. It can be shown from equation (10.4) that this signal/noise ratio is proportional to $(I_1 I_2)^{1/2}$, where I_1, I_2 are the light fluxes received by the two reflectors in that interval. The final r.m.s. residual

$$\sigma_r = [\sum_M (\overline{c_N(d)} \text{ (observed)} - \overline{c_N(d)} \text{ (computed)})_r^2 (I_1 I_2)_r / M]^{1/2} \qquad (11.6)$$

was then found for all M intervals and was minimized by optimizing the six free input parameters of the model in an iterative programme based on the Simplex method (Nelder and Mead, 1965).

The uncertainty in each of these optimum free parameters was found by computing the partial derivatives of the theoretical correlation with respect to each parameter at the time of each observation. These derivatives were then weighted in proportion to the signal/noise ratio, and the resulting matrix of weighted derivatives was multiplied by its transpose and inverted. The square root of the pth diagonal element of this inverse (x_{pp}) was then multiplied by the r.m.s. residual σ_r to find the r.m.s. uncertainty σ_p in the pth free parameter, where

$$\sigma_p = \sigma_r(x_{pp})^{1/2}. \tag{11.7}$$

Finally, to find the correct sense of rotation of the binary system, the whole analysis was carried out twice, for clockwise and anti-clockwise orbital motion. Although the two cases cannot be distinguished spectroscopically; they *can be distinguished* by the interferometer given sufficient observations of adequate signal/noise ratio, because the orbital motion adds to or subtracts from the parallactic angle. As a result, the correct sense of orbital motion gives a lower residual and, in the present case, it proved possible to distinguish it in this way.

11.4.4 *Results*

α Vir was observed on 12 nights in May 1966 with baselines of 10·0, 22·7, 59·7 and 88·3 m for a total time of about 84 hours. The effective wavelength used was $\lambda_0 = 443·0$ nm with a bandwidth of ± 5 nm. These results were analysed and yielded all the parameters of the orbit except the sense of orbital motion. The signal/noise ratio was not good enough to distinguish the correct sense from a simple comparison of the residuals σ_r; nevertheless, the difference between the inclinations (i) for the two solutions showed beyond question that the motion must be clockwise, because the 'anti-clockwise solution' yielded an inclination of 76° which implies eclipses which are not in fact observed.

Subsequently the signal/noise ratio of the interferometer was significantly improved, largely by the use of a new type of phototube, and it was decided to repeat the observations using a better choice of baselines. The second series were made on 16 nights in March and April 1970 with baselines of 19·7, 39·1 and 83·9 m for a total time of 115 hours. Again the effective wavelength was $\lambda_0 = 443$ nm with a bandwidth of ± 5 nm.

Analysis of this second set of observations yielded the unambiguous result that the sense of orbital motion is clockwise; the residual σ_r for clockwise rotation was significantly less than for anti-clockwise. The other parameters of the orbit were in satisfactory agreement with those found in 1966; as expected, the 1970 results were nearly twice as accurate as those for 1966.

The results of the observations in 1970 are shown in Table 11.5. Fig. 11.3 illustrates the variation of correlation with time observed with a baseline of 39·1 m. For the purpose of this figure we have made use of the fact that the orbital period of α Vir (4·014 days) is so close to 4 days that, over an observing period of 20 days, the binary system presents essentially the same phase every fourth night. Thus we have shown the observed correlation, as a function of hour angle, for

Parameter	Value ± r.m.s. uncertainty	Source†
Inclination of orbit (i)	$65°\cdot9 \pm 1°\cdot8$	I
Angular size of primary (θ_{UD1})	$(0''\cdot87 \pm 0''\cdot04) \times 10^{-3}$	I
Angular size of secondary (θ_{UD2})	$(0''\cdot4) \times 10^{-3}$	Assumed
Angular size of primary (limb-darkened) (θ_{LD1})	$(0''\cdot90 \pm 0''\cdot04) \times 10^{-3}$	I
Angular size of semi-major axis (θ_a)	$(1''\cdot54 \pm 0''\cdot05) \times 10^{-3}$	I
Brightness ratio of components (β)	$6\cdot4 \pm 1\cdot0$	I
Position angle of line of nodes (Ω)	$131°\cdot6 \pm 2°\cdot1$	I
Sense of orbital motion	Clockwise	I
Epoch of periastron passage (T)	JD 2440678·09	S
Eccentricity of orbit (e)	0·146	S
Longitude of line of apsides (ω)	138° at JD 2440678	S
Inverse period ($1/P$)	0·249091 days^{-1}	S
Period of rotation of line of apsides (U)	124 yr	S
Semi-major axis (a)	$(1\cdot93 \pm 0\cdot06) \times 10^7$ km	I+S
Distance	84 ± 4 pc	I+S
Mass of primary (m_1)	$10\cdot9 \pm 0\cdot9\ m_\odot$	I+S
Mass of secondary (m_2)	$6\cdot8 \pm 0\cdot7\ m_\odot$	I+S
Radius of primary (R_1)	$8\cdot1 \pm 0\cdot5\ R_\odot$	I+S
Surface gravity of primary ($\log g_1$)	$3\cdot7 \pm 0\cdot1$ [g_1 in c.g.s. units]	I+S
Absolute surface flux of primary ($\mathscr{F}_{v1}$ at $1\cdot83\ \mu^{-1}$)	$(2\cdot75 \pm 0\cdot24) \times 10^{-3}$ erg cm^{-2} s^{-1} Hz^{-1}	I+P
Effective temperature of primary ($T_{e1}(F)$)	22400 ± 1000 K	I+P
Luminosity of primary ($\log L_1/L_\odot$)	$4\cdot17 \pm 0\cdot10$	I+S+P
Absolute magnitude of primary (M_{V1})	$-3\cdot5 \pm 0\cdot1$	I+S+P
Absolute magnitude of secondary (M_{V2})	$-1\cdot5 \pm 0\cdot2$	I+S+P

† I = interferometric, S = spectroscopic, P = photometric.

Table 11.5. The parameters of Spica (α Vir). From Herbison-Evans, Hanbury Brown, Davis and Allen (1971).

12 nights divided into four sets of three nights each, each set corresponding to a particular phase of the binary system. The points show the average correlation observed for three nights with their associated r.m.s. uncertainties, and the full line shows the variation calculated for a binary star with the parameters given in Table 11.5.

There are a number of minor sources of error in this analysis which are discussed by Herbison-Evans *et al.* These include the effects of limb-darkening, distortion by rotation and tidal interaction, periodic changes in the primary star which is a β Cepheid variable, the reflection effect and so on. It is argued that, in the present context, none of these effects introduces significant error.

In considering the results of such a complex analysis, most of which takes place in a computer, one cannot help wondering if they are correct,

145

and to satisfy these doubts, Herbison-Evans *et al.* made a critical test. All the parameters of the orbit, except the period of rotation of the line of apsides which was taken as $U = \infty$, were left free and the computer was made to optimize all the 11 remaining parameters (θ_a, θ_{UD1}, θ_{UD2}, i, β, Ω, T, e, ω, C, and P) from the interferometer data alone; no spectroscopic data were used. The solution gave all the parameters, except θ_{UD2}, and again showed that the sense of orbital motion is clockwise; to determine θ_{UD2} would have required data from much longer baselines. The results for T, e, ω and P, based entirely on data from the interferometer, proved to be in satisfactory agreement with the spectroscopic values and gave striking confirmation that the whole analysis was correct.

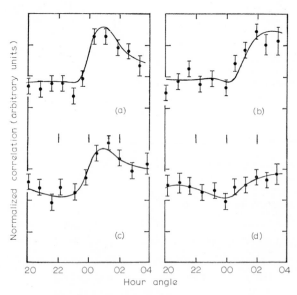

Fig. 11.3. The variation of correlation with hour angle for Spica (α Vir) for a baseline of 39·1 m. The observations were made on 12 nights in 1970 and have been grouped (as described in the text) to show four different phases. The points show the observations with their r.m.s. uncertainties; the full lines show the correlation calculated for a binary star with the parameters given in Table 11.7. From Herbison-Evans, Hanbury Brown, Davis and Allen (1971).

11.4.5 *Discussion of the results on Spica (α Vir)*

The interferometric results in Table 11.5 were combined with spectroscopic and photometric data to find some of the principal parameters of α Vir. The results are also shown in Table 11.5. As a first step the length of the semi-major axis (a) of the binary orbit was found by combining the spectroscopic data, which gave the projected axis ($a \sin i$), with the inclination (i) measured by the interferometer. The *distance* of the star was then found from the semi-major axis (a)

146

and the apparent angular size (θ_a) of the axis measured by the interfero-meter. The resulting distance is 84 ± 4 pc and the major contribution to the uncertainty in this result is, surprisingly, in the spectroscopic data.

The masses of the two stars were found by combining $m_1 \sin^3 i$ and $m_2 \sin^3 i$ found from the spectroscopic data with the inclination (i) given by the interferometer.

The angular size of the primary θ_{LD1} was found by correcting θ_{UD1} for limb-darkening, as described in § 10.3.2, taking a limb-darkening coefficient of 0.39. The radius of the primary R_1 was then found from θ_{LD1} and the distance; the surface gravity was found from the mass (m_1) and radius (R_1) of the primary. The effective temperature T_e and absolute surface flux of the primary F_λ were found by the procedure described in § 11.2.3 for single stars, and the luminosity L_1 was found from the radius and the effective temperature. Finally, the absolute magnitudes of the two stars (M_{V1}, M_{V2}) were found from the apparent magnitude of the binary ($V = 0.97$), the brightness ratio (β) of the two components and the distance.

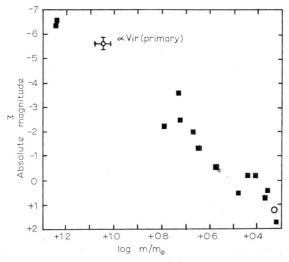

Fig. 11.4. Observational data on the mass-luminosity relation for early-type stars.

The results for α Vir shown in Table 11.5 are a striking demonstration of the value of a high resolution interferometer applied to the study of close binary stars. As we have seen, the measurements yield the inclination (i) of the orbit which enables us to find the masses from conventional spectroscopic data. The measurement of the angular size of the semi-major axis (θ_a) enables us to find the distance which, together with the luminosity ratio and angular size of the components, makes it possible to find the radius, surface gravity, temperature and

147

absolute magnitude of the stars. To illustrate the value of such measurements on early stars fig. 11.4 shows a plot of the existing data on the mass–luminosity relation for early stars. It is taken from the list of 'reliable' data given by Harris, Strand and Worley (1963) with the addition of the more recent data on CO Lac given by Smak (1967). The result on α Vir is shown by a cross and it is abundantly clear that more measurements would be valuable.

However, the most interesting aspect of this work on α Vir is the measurement of distance. If we look in the *General Catalogue of Trigonometrical Parallaxes* (Jenkins, 1963) the distance of α Vir is given as 53 pc, which is the weighted mean of three independent measurements giving 111, 59 and 34 pc. This large scatter is, of course, to be expected because at such a large distance the uncertainties in the trigonometrical parallax are of the order of ± 50 per cent. Thus, this measurement represents an extension of the classical method of measuring parallax by, very roughly, a factor of 10. It must also be noted that the result depends only on velocity and angle and therefore, like the classical measurements of parallax, is independent of interstellar extinction or spectroscopic criteria of luminosity. The possibilities of such a technique are of considerable interest and are discussed further in § 12.3.

11.5   The emission-line star γ Velorum

A surprising feature of equation (4.30) is that the signal/noise ratio of an intensity interferometer is independent of the optical bandwidth, provided only that this bandwidth is much greater than the post-detector electrical bandwidth. The electrical bandwidth of the Narrabri interferometer was about 100 MHz and so, at 443 nm, the optical bandwidth could in principle be reduced to about 10^{-4} nm without loss of signal/noise ratio. In practice, of course, such extremely narrow bandwidths could never have been used because it would have been impossible to match the two optical filters with the necessary precision, nor could they have been made to operate in the very poorly collimated beam from the reflectors. Nevertheless, it was possible to use filters with significantly narrower bandwidths than the ± 5 nm used in the main programme. This possibility suggested that we could measure the apparent angular size of a star in both the light of an emission line and the continuum so that the size and shape of the emission region, relative to the exciting star, might be found directly.

The only suitable star, sufficiently bright and within view of the interferometer, was γ Velorum. This star consists of three components, the relatively faint star γ^1 Vel and the bright spectroscopic binary γ^2 Vel. The binary shows strong emission lines of ionized carbon and the components have been classified as WC–8 (Wolf–Rayet) and O7; these two stars differ in brightness by about one

148

magnitude and it was believed (e.g. Smith, 1968) that the Wolf–Rayet is the brighter. More recently a detailed spectroscopic analysis (Conti and Smith, 1972) points to the conclusion that the O7 star is the brighter.

Observations of γ Vel were made at Narrabri in 1968 and have been reported by Hanbury Brown, Davis, Herbison-Evans and Allen (1970). The star was observed in the continuum at 443 nm through the standard interference filters which have a bandwidth of 10 nm; altogether six baselines, ranging from 10 to 188 m were used and the total exposure time was about 80 hours. Observations were then made in the light of the C III–IV emission line at 465 nm through filters with a bandwidth of 2·5 nm; four baselines, ranging from 10 to 56 m were used and the total exposure was about 50 hours.

The results showed that the correlation in the emission line (465 nm) decreased more rapidly with baseline than in the continuum (443 nm). It follows, as expected, that the angular size of the system is much greater in the emission line. A detailed analysis, greatly complicated by the difficulties of separating continuum and emission line, has been given by Hanbury Brown et al. They found that the angular diameter, averaged over a range of position angles, is $(2·05 \pm 0·19) \times 10^{-3}$ seconds of arc in the emission line and $(0·44 \pm 0·05) \times 10^{-3}$ seconds of arc in the continuum. Thus the apparent size of the emission region is roughly five times greater than that of the brighter star. Also, some rough information about the shape of this region was extracted from the variation of correlation with hour angle or position angle; this showed that the emission region cannot be very asymmetrical in shape because the ratio of maximum to minimum angular size was less than $\frac{3}{2}$ over a wide range of position angles. The surface flux (F_λ) from the emission region was found to be $(1·17 \pm 0·22) \times 10^{-7}$ erg m^{-2} s^{-1} Hz^{-1} at 465 nm corresponding to a brightness temperature $T(465 \text{ nm}) = 12600 \pm 900$ K. The effective temperature of the brighter of the two stars in γ^2 Vel was found to be $T_e = 30\,100 \pm 4000$ K.

The remainder of the analysis by Hanbury Brown et al. is concerned with the orbital parameters of the binary, distance, radii, masses, etc. and we shall not review it here. It suffices to say that this experiment demonstrated that an intensity interferometer can be used to measure the angular size and brightness, possibly also the shape, of an emission region. It would be of considerable interest, given a more sensitive instrument, to observe other emission-line stars and particularly some Be stars. This point is referred to briefly in § 12.5.

11.6 *Limb-darkening of Sirius*

In § 10.3.2 we noted that any significant difference between the variation of correlation with baseline for stars showing different degrees of limb-darkening only appears in the secondary maximum of the curve. This point is illustrated in fig. 10.2 which shows that the amplitude of this secondary maximum is very small and is expected to be of the order 10^{-2} of the zero-baseline correlation. If we are to make a significant measurement of limb-darkening, then it is clear that the signal/noise ratio must be high and we are restricted to very bright stars. At Narrabri the only possibilities were α C Ma and α Car; their signal/noise ratios were such that in a measurement of correlation lasting about 25 hours the uncertainty in the result corresponded to about 2×10^{-3} of the zero-baseline correlation. Although this sensitivity was marginal we decided that it would be worth while to try to measure the limb-darkening of Sirius. Even if we failed to gain any useful astrophysical data it would be valuable to the design of any future instrument to explore the practical difficulties.

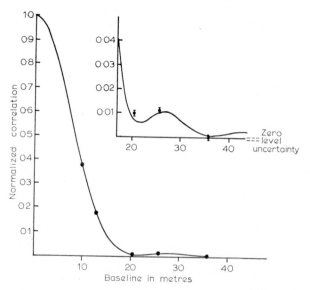

Fig. 11.5. Correlation as a function of baseline for Sirius A (α C Ma). The points show the observed results; the full line shows the theoretical curve for a model atmosphere ($T_e = 10\,000$ K, $\log g = 4$, $\lambda = 450$ nm). Results for three long baselines are shown on an expanded scale together with their r.m.s. uncertainties. (Total exposure 203 hours.)

We observed Sirius at five different baselines in 1969, 1970 and 1971 for a total period of 203 hours. The results, normalized to unity at zero-baseline, are shown in fig. 11.5. It is interesting to note that the amplitude of the secondary maximum is roughly 10^{-2}, as expected; presumably this is the first occasion on which it has ever been measured for a star.

To compare these results with theory we have not used the simple cosine approximation discussed in § 10.3.2. We have used the radial distribution of brightness across the disc given by Gingerich (1969)

150

for a model stellar atmosphere with $T_e = 10\,000$ K, $\log g = 4$ and $\lambda = 450$ nm. The angular diameter of the model was adjusted to give the best fit to the observed points and the calculation took into account the partial resolution of the disc by the reflectors. The results are shown by the full line in fig. 11.5. It can be seen that the observations are in reasonable agreement with theory.

There was one peculiar point about these observations which may be worth recording. An analysis of the correlation observed at the three longer baselines suggests that it depended on the position angle of the star. The changes with position angle were barely significant and it is difficult to decide whether they were real or not and whether they were associated with the star or represented some minor systematic error in the equipment. As far as the equipment is concerned, it is difficult to trace an effect which amounts to only about 5×10^{-3} of the zero-baseline correlation; nevertheless, we explored all the systematic errors which we could imagine to vary with the position angle of the star. These included the effects of elevation angle on the optical bandwidth and hence of the normalized correlation, unwanted coupling between the arms of the interferometer as a function of the azimuth of a star, radio interference as a function of azimuth or time, systematic temperature changes throughout the night and so on. No explanation could be found.

We also explored the possibility that these small changes in correlation were due to the star. The first and most obvious suggestion is that Sirius A is a double star with an angular separation less than 0·1 seconds of arc. The suggestion that Sirius A may be double has been made by other authors (e.g. Heintze, 1968, Lindenblad, 1970) on the evidence of its spectrum and of very small irregularities in its motion. We could find, however, no evidence for a companion in the published radial velocities of Sirius A and, although a model can perhaps be contrived to fit all the observations, it remains unconvincing unless there is some other supporting evidence. Other possibilities exist and one of the most attractive is to imagine Sirius A to be surrounded by, or close to, some shell or body of gas. We are left in the unsatisfactory position of suspecting that there may be something odd about Sirius A but we cannot be sure because all the evidence is marginally significant.

11.7 *The Rotation of Altair*

The effects of rotation on the apparent angular size of a star are discussed in § 10.3.4. It is concluded that for most stars these effects are so small that they would have been difficult to detect with the Narrabri interferometer. However, we decided to try to explore the effects of rotation on the bright and rapidly rotating star Altair (α Aql).

The main problem in making measurements was to observe the star over a sufficiently wide range of position angles. Thus if Altair is observed for 3 hours before and after transit, corresponding to elevations above 30°, the total change in position angle seen by the interferometer is only 92° which is insufficient for a satisfactory experiment. We extended this range by arranging that observations could also be made with the baseline parallel to the direction of the star. In this mode the reflectors were set to follow the computed azimuth of the star $+90°$, and $-90°$ was added to their turntables in the appropriate direction so that they looked, one over the other, at the star. Their separation was controlled, in small steps, throughout the observation so that the projected baseline, as seen by the star, remained constant. The delay in the light reaching the more distant reflector was compensated at the input to the correlator by a variable delay network.

Most of the measurements were made with a baseline of 18 m, which is the distance at which the correlation has fallen to about one-half of that at zero-baseline. The star was observed in 1970 and again in 1971 for a total of 84 hours with a baseline normal to the direction of the star and 56 hours with a projected baseline parallel to the direction of the star. The normalized correlation observed at these two baselines is shown in fig. 11.6 as a function of the position angle of the star. The correlation shows a significant variation with position angle and the broken line shows the best fit of a sinusoid to the data. The amplitude of this variation is 9 ($\pm$6) per cent of the normalized correlation at 18 m.

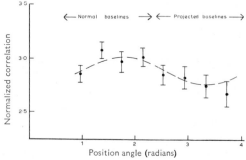

Fig. 11.6. Correlation as a function of position angle for Altair (α Aql) observed with a baseline of 18 m. (140 hours' exposure.)

It is interesting to compare these data with theory. We have already noted in § 10.3.4 that Johnston and Wareing (1970) predicted a maximum variation of 4 per cent in the apparent angular size and it is clear that our observations show a larger effect. But in their calculations they took the equatorial velocity of Altair to be about 0·8 of the break-up velocity and there are good reasons (e.g. Hardorp and Strittmatter, 1968, Hardorp and Scholz, 1971) to believe that this value is significantly too low. We have therefore compared our results with the rotating models of Hardorp and Strittmatter. Dr. Strittmatter kindly supplied us with the details of a model of Altair assuming an equatorial velocity of 0·99 of break-up. For this model we calculated the variation of correlation with baseline as a function of inclination and position angle. Fig. 11.7 shows the results for zero inclination and for two position angles at right-angles, corresponding to the axis of rotation parallel and normal to the baseline of the interferometer. The difference between these two curves gives the maximum variation of correlation with position angle and at a baseline of 18 m ($\pi d\theta/\lambda = 2\cdot1$) it is about 16 per cent, somewhat larger than the observations indicate.

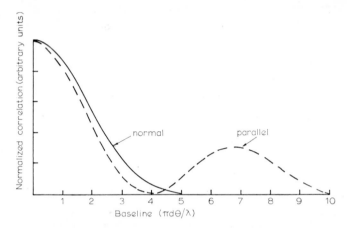

Fig. 11.7. Correlation as a function of baseline for a theoretical model of Altair (α Aql). The baseline is in units of ($\pi d\theta/\lambda$) where θ is the maximum (equatorial) angular diameter and d is the length of the baseline. The full line is for a baseline normal, and the broken line for a baseline parallel, to the axis of rotation of the star.

Although it is interesting that the expected variation of correlation does take place, the signal/noise ratio is clearly not good enough to allow any critical comparisons. The uncertainties in the results would allow a variety of models and a wide range of angles of inclination of the star's axis of rotation.

As a further and more detailed check on Strittmatter's model it would be valuable to observe the large variation of correlation predicted at the baseline corresponding to the secondary maxima of the curves in fig. 11.7. Unfortunately we only made measurements at the appropriate distance with the baseline normal to the direction of the star; no correlation was seen. Subsequent analysis of the data taken at 18 m indicates that the expected peak in correlation at this longer baseline would be of comparatively shorter duration and would be observed at a position angle covered only by the projected baselines. This would be an interesting experiment to do, preferably with a more sensitive instrument.

153

11.8 Polarization Tests on Rigel

In §10.3.5 we pointed out that the apparent angular diameter of a star may depend upon the plane of polarization of the light accepted by the interferometer. We estimated that electron scattering in a stellar atmosphere is unlikely to produce a change as large as 1 per cent; but there is the possibility that electron scattering in an ionized corona, possibly associated with mass loss from a hot star, might produce larger effects which could be measured. With this in mind we decided to observe the supergiant B8 star Rigel (β Ori) in two orthogonal planes of polarization.

The optical system of the interferometer was modified to incorporate two polarizers which were mounted in front of the phototubes. These polarizers were remotely controlled from the control desk so that they could be rotated into either of two orthogonal directions, parallel or normal to the baseline. Care was taken to ensure that corresponding planes of polarization were parallel in the two reflectors. Measurements showed that at 443 nm the polarizers passed about 40 per cent of the incident light.

Rigel was observed during November and December 1971 for a total time of 19 hours at a baseline of 9·95 m and 59 hours at 19·68 m. The observations were carried out by the normal procedure described in §9.5 except that the polarizers were rotated through 90° every ten cycles (1000 s) of the correlator. The two sets of observations were therefore interleaved in time, but were analysed separately as completely independent measurements of the angular diameter of the star. If θ_H, θ_V represent the apparent angular size of the star with the plane of polarization (electric vector) parallel and normal to the baseline respectively, then we observed that

$$\theta_H = (2\cdot38 \pm 0\cdot07) \times 10^{-3} \text{ seconds of arc}$$
$$\theta_V = (2\cdot44 \pm 0\cdot06) \times 10^{-3} \text{ seconds of arc}$$
$$\theta_V/\theta_H = 1\cdot025 \pm 0\cdot040$$

Thus we observed no significant difference between the apparent angular diameters of Rigel in the two planes of polarization.

We have used this result to find an upper limit to mass loss from Rigel using the very simple model outlined in §10.3.5. Taking the upper limit $\theta_V/\theta_H < 1\cdot065$ it can be shown that the corresponding limit to the scattering parameter x in equation (10.19) is, approximately, $x < 0\cdot25$. Putting $R = 50$ solar radii, from equation (10.19), the ratio of mass to velocity $S/V < 40$. There is no spectroscopic evidence for mass loss from Rigel but if we put $V = 500$ km s^{-1} (the escape velocity), then the mass loss S must be less than 2×10^{-5} $M\odot$ yr^{-1}. Although this conclusion is itself of little astrophysical interest the experiment points the way to a new way of observing an ionized corona around hot stars which might well prove to be valuable given a more sensitive interferometer.

11.9 Observations of Čerenkov Light Pulses

As we have already discussed in §5.7, Čerenkov light pulses due to cosmic rays entering the Earth's atmosphere will produce correlated signals in two separated light detectors.   If this correlation is significant, compared with that from the star under observation, errors may be introduced into the measurements of angular size.   Early estimates by Hanbury Brown and Twiss (1958 a) suggested that this unwanted correlation would be negligibly small in the interferometer at Narrabri. Nevertheless, their estimates were necessarily uncertain and so the observations described here (Hanbury Brown, Davis and Allen (1969)) were made to put these estimates on a more satisfactory basis; they were not intended to be a general investigation of Čerenkov radiation.

For the purpose of these experiments it was necessary to enhance the effects of Čerenkov radiation in the two reflectors by increasing their optical bandwidth and angular field of view.   The field was increased from 20 to 34 minutes of arc by using larger phototubes and by removing the optical system shown in fig. 8.3, thereby exposing the phototubes to white light.   The electronic correlator was replaced by a discriminator and pulse counter in each channel and by a coincidence counter between the two channels.

The absolute rate of Čerenkov light pulses was found from

$$n_{\mathrm{cv}} = n_{\mathrm{c}} - n_r \qquad (11.8)$$

where n_{c} was the observed coincidence rate and n_r was the random coincidence rate.   The Čerenkov rate n_{cv} was measured with the two reflectors separated by 10 m and pointing at the same region of sky; at $55°$ elevation it was found that the rate was

$$n_{\mathrm{cv}}(>h) \approx 6(h/h_0)^{-1} \qquad (11.9)$$

where h is the pulse height and h_0 the height of a pulse produced by a single photoelectron.   This counting rate increased with elevation and if, following Jelley and Galbraith (1955), it is assumed that

$$n_{\mathrm{cv}}(z) \propto \cos^l z \qquad (11.10)$$

where z is the zenith angle, then measurements at Narrabri made at two different elevations gave $l = 3{\cdot}0 \pm 0{\cdot}3$.

Experiments were also carried out to find how the rate of coincident pulses n_{cv} depended upon the separation between the reflectors and their relative alignment.   The two reflectors were fixed at the same elevation ($39°$) and pointed (± 1 minute of arc) in the same direction. The rate was then measured for 300 s.   The reflectors were then rotated on their turntables through equal but opposite angles so that their baseline and elevation angles remained constant but their pointing directions differed in azimuth by an angle θ.   The rate was recorded for

155

several values of θ; these measurements were carried out at three different baselines, 10, 49 and 94 m.

The results are shown in fig. 11.8 where the angle θ is defined as positive when the two reflectors were turned towards each other. They show that the counting rate increased as the reflectors were turned towards each other and reached a maximum at an angle which increased with baseline. It is interesting to note that both the maximum counting rate and the apparent angular width of the distribution was independent of baseline length over the range 10–94 m. It follows that the Čerenkov light pulses seen at all three baselines appear to come from a source about 8·4 km above mean sea level and with a finite angular size of roughly 0·75° between points of half intensity.

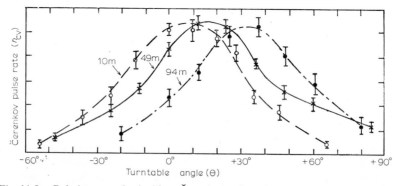

Fig. 11.8. Relative rate of coincident Čerenkov pulses observed in two spaced reflectors as a function of their alignment. The angle of elevation was 39° and the results are shown for three different baselines. θ is the misalignment of the two reflectors in azimuth and is positive when they are turned towards each other. From Hanbury Brown, Davis and Allen (1969).

In their account of these observations Hanbury Brown, Davis and Allen (1969) estimate that the Čerenkov light pulses which they detected were due to cosmic rays with a primary energy exceeding 10^{12} eV. They compare the results with the theoretical work of Zatsepin (1964) and conclude that they are in reasonable agreement.

These data put the estimates of the effects of Čerenkov radiation on an intensity interferometer on a firm quantitative basis. They show that any correlation due to this source must vary with zenith angle, baseline and relative misalignment of the two reflectors. They are used in § 5.7 to show that this correlation, which is a potential source of error in measuring stars, is almost certainly negligible in the Narrabri stellar interferometer. Furthermore, it is shown in § 5.7 that, in any more sensitive interferometer designed to reach fainter stars, the effects of Čerenkov radiation are likely to be negligible.

11.10   A Search for Sources of False Correlation

One of the possibilities which we had clearly in mind throughout the observations at Narrabri was the chance that there might be sources of correlation other than the star under observation.   Any such source might perhaps vary with baseline or time in such a way as to alter the true ratio of correlations observed at different baselines and thereby falsify the measurements of angular diameter.   At the start of our programme we envisaged three possible sources of false correlation— Čerenkov light pulses from the night sky due to cosmic rays, radio interference picked up on both arms of the interferometer and unwanted electrical coupling between the two arms.   Later in the programme we added a fourth when we noticed a report by Morris (1971) that he had observed the modulation of the light from several stars by discrete radio-frequencies.   He claimed that the modulation percentage was about 1 per cent and that most of the frequencies lay in the range 3 to 8 MHz.   He attributed this modulation to scattering of the starlight by enhanced electron density fluctuations in the ionosphere.   We realized that such an effect, if it exists, might produce spurious correlation which presumably would vary both with baseline and time.

As a check on these four specific sources we made a number of special tests at Narrabri.   The measurements of Čerenkov light pulses have already been described in § 11.9.   They put the estimates of correlation on a sound quantitative basis and showed that any unwanted correlation due to these pulses is negligible.

As a check on radio interference a search was made on several occasions for radio signals in the output of the main amplifier of the correlator using a narrow-band communication receiver and, on one occasion, also a spectrum analyser; measurable signals could only be found when there was known to be something wrong with the double screen of the cables from the correlator to the phototubes.

As a check on unwanted electrical coupling between the two arms of the interferometer, prior to the correlator, stringent tests were carried out by injecting a strong signal into one channel and using the correlator to detect the presence of a coupled signal in the other channel.   These experiments were carried out in both channels, first with the reflectors close together in the garage and then with them widely spaced on the track.   These tests were extremely sensitive and it was gratifying to find that no significant coupling could be measured. It is worth noting that any electrical coupling which takes place before the phase-switches can produce false correlation in the same way as a true correlated source; however, if the coupling takes place after either of the phase-switches then false correlation can only be produced by second-order processes in the correlator for which the sensitivity is greatly reduced.   It is, of course, for this reason that one of the phase-switches was mounted close to the focus of one reflector so that the effects of any coupling between the long cables would be greatly reduced.

157

As a check on the radio-frequency modulation of starlight one reflector of the interferometer was directed at the bright star α Lyr and then at β Cen. The spectral distribution of the output noise from the main amplifier of the correlator was examined in detail with a spectrum analyser (Hewlett Packard type 8554). The analyser was set to have a radio-frequency bandwidth of 1 kHz and an output bandwidth of 100 Hz; it was scanned slowly over the range 0–10 MHz in two steps and the output was recorded by a pen recorder. Several scans of the spectrum were made and compared immediately with scans taken with the phototube illuminated by a small pea-lamp giving the same light flux as the star. A few scans were made over the range 0–100 MHz. No trace of any radio-frequency modulation of the starlight was detected and the sensitivity was such that a modulation depth of about 0·5 per cent would have been seen. Obviously these tests were not exhaustive as they were conducted on only two nights (20 and 21 June 1972); nevertheless, the experimental results offered by Morris (1971) were not, in our opinion, sufficiently convincing to justify a more extensive search.

As a final overall check that there were no significant sources of false correlation due to any cause, suspected or unsuspected, the following tests were carried out. The star β Cru was observed, following the standard procedure, but with a very long baseline of 154 m. At this baseline the star must be completely resolved and it was therefore expected to produce zero correlation. In an exposure of 55 hours, spread over many nights, no correlation was observed and this result implies that at 154 m any false correlation must have been less than 2·5 per cent of the zero-baseline correlation $c_N(0)$ expected from an unresolved star of magnitude $+1·0$. Two further tests were then made with the shortest possible base-line (10 m). In the first of these tests the reflectors followed a region of sky with no bright stars over a wide range of elevation angles and, in an exposure of 7 hours no correlation was observed. This test was repeated for 10 hours with the reflectors at a fixed elevation of $52°$; again no correlation was observed. Both of these tests were carried out with the system at full gain and, because there was no noise in the correlator output due to starlight, they represent very sensitive tests for correlation due to Čerenkov light, radio interference or any other source in the night sky.

The results of these latter observations are shown in fig. 5.2 as experimental upper limits due to Čerenkov light. They were also used, together with the tests on coupling, to estimate the upper limits to any false correlation shown in fig. 10.1. These limits depend upon the gain of the phototubes and the correlator and so they vary with the brightness of the star under observation as shown in the figure. The limits in fig. 10.1 have been used, as discussed in § 10.2, to estimate the uncertainty in the final measured values of angular diameter.

11.11   *The Effects of Atmospheric Scintillation*

In § 5.8 we presented theoretical arguments to show that the effects of atmospheric scintillation on the measured values of correlation should be negligibly small.   In an attempt to verify this experimentally we made two tests.   Firstly, on the assumption that the effects of scintillation are likely to vary from night to night, we compared the dispersion in the normalized correlation $(\overline{c_N(d)} \pm \sigma_N)$ observed from stars on 126 different nights with the dispersion to be expected, see equation (10.8), in the correlator output due simply to noise and zero-drift; the magnitude of the expected fluctuation was established experimentally by dummy runs during which the phototubes were illuminated by lamps.   For 126 nights, the dispersion in $\overline{c_N(d)}$ was 0.99 ± 0.06 of the expected value.   There was therefore no significant increase in dispersion due to scintillation.

In a second test we measured the normalized correlation from the bright star Sirius as a function of elevation.   At the same time we recorded the peak to peak fluctuations in the output current of one phototube as a measure of the scintillation.   The results from a single night are shown in fig. 11.9.   It can be seen that, although the scintillation increased markedly at low angles of elevation, there was no corresponding decrease in the normalized correlation.   This result, which was confirmed on other nights, points to the conclusion that any variation of normalized correlation with scintillation must have been small.

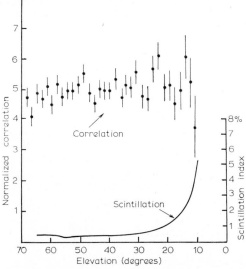

Fig. 11.9.   Normalized correlation as a function of elevation angle observed for Sirius (α C Ma).   The points show the observed correlation with its r.m.s. uncertainty; the full line is the observed scintillation index expressed as the peak to peak percentages fluctuation in the phototube anode currents.

Admittedly these two tests were not exhaustive; in particular they were not satisfactorily precise and they are also open to the criticism that no conventional estimates were made of the quality of the seeing on the various nights. Nevertheless, they were adequate for our immediate purpose and offer strong support to the theoretical prediction that the effects of scintillation are negligibly small. We may conclude that scintillation is unlikely to have introduced significant errors into the measurement of angular diameter. It must be remembered that even if there were small effects due to scintillation, it is unlikely that they introduced errors into the angular size because observations at each baseline were always carried out over the same range of elevations.

11.12 Signal-to-Noise Ratios

A theoretical expression for the signal/noise ratio of an intensity interferometer is given in equation (5.17) which is reproduced here for convenience:

$$(S/N)_{\mathrm{RMS}} = (A_1 A_2)^{1/2} \alpha(\nu_0) g(\nu_0) n(\nu_0) \Delta(\nu_0) \Gamma^2(\nu_0, d) \epsilon \beta_0 \sigma$$
$$((\mu - 1)/\mu) (b_{\mathrm{v}} T_0 / 2\eta)^{1/2} / (1 + a)(1 + \delta). \qquad (11.11)$$

For the interferometer at Narrabri during the years 1970 and 1971 the parameters in this equation were estimated to have the following values: area of reflectors $(A_1 A_2)^{1/2} = 29 \cdot 5$ m; quantum efficiency at 443 nm averaged over photocathode, $\alpha(\nu_0) = 0 \cdot 20$; overall transmittance of optical system (mirrors, lenses, filters), $g(\nu_0) = 0 \cdot 40$; efficiency of correlator, $\epsilon = 0 \cdot 90$; polarization factor, $\beta_0 = 1$; optical spectral density factor (equation (5.6)), $\sigma = 0 \cdot 85$; excess noise in phototubes, $(\mu - 1)/\mu = 0 \cdot 95$; bandwidth of electronic system including phototubes, $b_{\mathrm{v}} = 55$ MHz; spectral density factor of the electronic system (equation (5.16)), $\eta = 0 \cdot 75$; stray light and dark current, $(1 + a) = 1 \cdot 01$; excess noise in correlator, $(1 + \delta) = 1 \cdot 10$. Taking $n(\nu_0) = 0 \cdot 95 \times 10^{-4}$ photons $\mathrm{m}^{-2} \mathrm{s}^{-1} \mathrm{Hz}^{-1}$ at 443 nm as the flux from a star of zero magnitude $(B = 0)$ at the top of the Earth's atmosphere and the atmospheric absorption to be $0 \cdot 39$ mag/air mass, then equation (11.11) gives the signal/noise ratio from an unresolved $(\Delta(\nu_0)\Gamma^2(\nu_0, d) \approx 1)$ star of magnitude B, seen at 45° elevation as

$$(S/N)_{\mathrm{RMS}} = 0 \cdot 53 (T_0)^{1/2} 10^{-0.4B} \qquad (11.12)$$

where T_0 is the time of observation in seconds.

Fig. 11.10 shows a comparison between the ratios given by equation (11.12), shown as a full line, and those actually measured at Narrabri for five stars in the period 1970–71. These ratios were all measured at the shortest possible baseline (10 m) where these five stars were not significantly resolved. It can be seen that the observed ratios were roughly three-quarters of the theoretical values. No definite explanation of this discrepancy can be given; it is possibly due to the neglect of some factor in the theoretical analysis but it is more likely to be the cumulative result of a number of small errors in the parameters we have substituted in equation (11.11). For example, the values of $g(\nu_0)$ the overall loss in the optical system, ϵ the efficiency of the correlator and $(\mu-1)/\mu$, the excess noise in the phototubes were all estimated because they were too difficult to measure with satisfactory precision and in each case the estimate is likely to be optimistic; furthermore, the observations were carried out on many different nights and the average atmospheric absorption was likely to have been somewhat higher than the assumed value of 0·39 mag/air mass which corresponds

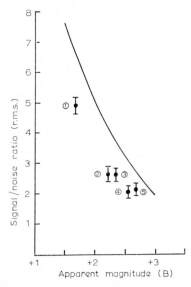

Fig. 11.10. Signal to noise ratio versus apparent magnitude for an unresolved star. The signal to noise ratio (r.m.s.) observed in an exposure of 1 hour for (1) β Car, (2) α Oph, (3) η C Ma, (4) ζ Oph, (5) γ Crv. The full line is the theoretical value from equation (11.12).

to a very clear night. As a result the theoretical signal/noise ratios represent an upper limit and, due to the neglect of minor losses, it is to be expected that the observed ratios will always fall below this limit. In the present case the discrepancies shown by fig. 11.10 are comparable with the uncertainty in the theoretical values and we may conclude that the signal/noise ratios observed at Narrabri are consistent with the theoretical analysis in chapter 5.

161

M

CHAPTER 12

future possibilities

12.1 *Introduction*

We are now in the position to foresee how the work of Narrabri Observatory could be developed by building a more sensitive interferometer. After ten years' work the existing instrument has reached the limit of its capabilities and it is not practicable to improve it by the amount required to explore the many exciting possibilities which demand a higher sensitivity. We shall have to think in terms of an entirely new instrument.

We must first ask whether or not intensity interferometry is a fruitful technique to develop. It is true that, in principle, Michelson's interferometer should give a superior sensitivity but, as we have pointed out in § 4.1, its development has so far been limited by the difficulties of achieving the necessary mechanical precision in a larger instrument and of making precise measurements in the presence of atmospheric scintillation. It may well be that in the course of the next few years the problems of making an improved version of Michelson's interferometer at optical wavelengths with baselines of a few metres, or even tens of metres, will be solved but this would still limit its application to the cool stars only. Another interesting possibility is the development of coherent 'heterodyne' interferometers using lasers; these are strictly analogous to radio interferometers. Such interferometers require that the phase of the wave over the reflector should be uniform and so it seems that their principal application will not lie in the optical band where the wavefronts are distorted by atmospheric scintillation but in the far infra-red where wavefronts are more likely to be plane and where measurements should have valuable application to cool stars and other infrared sources. Thus for measurements of *hot stars* in the visible spectrum and near infra-red, which require baselines of hundreds or thousands of metres, it seems unlikely that an instrument superior to the intensity interferometer will emerge in the near future; at present it would appear that intensity interferometry is the only technique which can be used to measure the angular diameters of a wide range of stars with precision through the Earth's atmosphere and, for hot stars at least, it is likely to remain so for a long time.

Let us then consider some of the immediate programmes which a more sensitive intensity interferometer might tackle, bearing in mind that the most important results of research may well prove to be those which one cannot foresee. We shall also discuss briefly how the necessary sensitivity could be obtained and what form the new instrument might take.

12.2   *The Measurement of Emergent Fluxes, Effective Temperatures and Radii of Single Stars*

A major aim of a more sensitive interferometer would be to extend our knowledge of the fundamental quantities F_λ the absolute surface flux and T_e the effective temperature of single stars.   Our existing empirical data on these quantities are limited by the few angular diameters of single stars which have been measured.   Fig. 12.1 shows a spectral type-luminosity array of the single stars which have been measured at Narrabri.   It also shows the very few stars, all cool giants or super-giants, which have been measured with Michelson's interferometer or by lunar occultations.

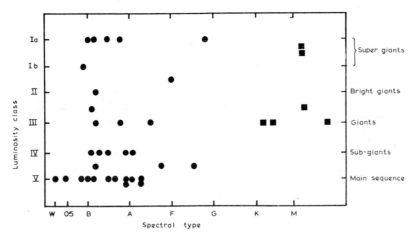

Fig. 12.1.   Stars with measured angular diameters.   ●: measured with intensity interferometer at Narrabri, ■: measured with Michelson's interferometer.

It can be seen at a glance that the existing data are sparse over the whole diagram and, in particular, there are almost no reliable data for stars cooler than type F, or hotter than type O 9·5. The one very hot type O star (ζ Pup, O 5f) measured at Narrabri is probably not typical of type O stars in general.

A basic contribution to stellar astronomy could therefore be made by an interferometer which is sufficiently sensitive to extend the measurements to many more stars so that F_λ and T_e could be found for a wider range of spectral types and luminosity classes and also as a function of age, metal content, spin, etc.   A simple count shows that there are very few bright hot stars and if we seek to measure a reasonable

163

sample of the extremely hot stars we must build an instrument which reaches to a visual magnitude (V) of at least $+7$; for example, for an interferometer sited at latitude $30°$ S, there are no type O 5 stars brighter than $+5$ and only four brighter than $+7·3$. At the other end of the spectrum—M stars—we need an instrument of about the same sensitivity but capable of working in the near infra-red. Thus to reach a main sequence star of type M 5 we must work at about 800 nm. Between these two extremes, O 5 and M 5, there will be a very large number of stars within reach of such an instrument; in the five spectral categories (B, A, F, G, K) there are about 15 000 stars brighter than $+7·3$.

Another application of a more sensitive interferometer would be to find the radii of single stars as a function of spectral type, luminosity class, etc. Obviously the most direct method is to measure the angular diameters of stars with known parallax. Since the bright early-type stars are too far away for accurate measurements of parallax, this method would be confined in practice to stars in the range type A 0 to M 1. For example, if we limit the work to stars with an uncertainty of less than 10 per cent in their parallax, then our more sensitive interferometer (limiting visual magnitude $+7$) would be able to measure only about 28 stars, all of these being in the spectral range A 0 to M 1. Although this number is disappointingly low, it must be remembered that the number of stellar radii which have been measured is small—less than 20—and most of them were derived from the study of eclipsing binaries.

For stars earlier than type A 0 there are two possible ways of finding the radius. It should be possible to measure their angular sizes in clusters of known distance; alternatively, the radii of a few hot stars in spectroscopic binaries might be found if their distance can be measured by the method discussed in § 11.4.

12.3 *Double Stars*

The observations of α Vir described in § 11.4 show how an interferometer can be used to find important parameters of a close binary star which cannot be determined spectroscopically. Thus, one can find the angular size of the semi-major axis, the inclination of the orbit, the angular diameter of at least the brighter component and the brightness ratio of the two components. These results, when combined with spectroscopic observations of a double-lined binary, yield the distance, masses and absolute magnitudes and luminosities of the two stars and the radius of the primary.

In view of the lack of data about the masses and luminosities of early-type stars, this is clearly an application of considerable importance and it is worth taking a quick look at the number of binary stars which could be measured. The first point to consider is that binaries cannot be measured if the angular separation of the two components is too great and would be resolved by the individual reflectors of the interferometer. In effect this sets an upper limit to the period of a binary as a function of the reflector size, the wavelength and the brightness and spectral types of the stars. If we exclude all those stars with too great a separation, then a count of the remaining double-lined spectroscopic binaries shows that there are 35 systems which could probably be measured with a more sensitive interferometer sited at 30° S and with a limiting magnitude of +7 on single stars. In making this estimate we have taken the limiting magnitude on a binary star to be +6·0, one magnitude brighter than a single star, because detailed measurements of the parameters of a binary demand a higher signal/noise ratio.

By observing these binaries one could measure some reasonably precise distances, independent of interstellar extinction, out to distances as great as 1 kpc. It may therefore be possible to measure directly the distances of one or two clusters or associations. However, it seems likely that the main value of this work would be to establish the masses, luminosities and radii of early-type stars.

There is little doubt that the detailed observations of double-lined spectroscopic binaries would be the major double-star programme of a more sensitive interferometer but it might also make an interesting contribution to the general study of binaries because it can distinguish between single and double stars. At Narrabri this can be done with reasonable certainty provided that the two components of a double star differ by less than 2·5 magnitudes. In an improved instrument it should be possible to increase this limit by perhaps one magnitude for bright stars. Experience suggests that many previously unsuspected double stars would be found and that these data would be valuable both in the individual cases and in the general statistics of double stars.

12.4 *Cepheid Variables*

One of the most attractive possibilities of a stellar interferometer is to study Cepheid variables. It is disappointing that the existing interferometer is not sufficiently sensitive to engage in this work and that the necessary modifications appear to be unreasonably extensive. From latitude 30° S one could observe seven classical Cepheids brighter than visual magnitude +5 at minimum light, nine brighter than +6 and 20 brighter than +7. If we consider an interferometer with a limiting magnitude of +7 on single stars, then, allowing for the necessary signal/noise ratio, there seems to be no reason why we should not measure the changes in apparent angular size as a Cepheid pulsates. The results, combined with photometric data, would give the surface flux and temperature as a function of phase. Furthermore, by combining the observed change in angular size with spectroscopic measurements of the radial velocity of pulsation it appears possible, after applying certain theoretical corrections for changes in the atmospheric depth, to find the distance and radius of the star. On the very brightest Cepheids (e.g. η Aql, ζ Gem, ρ Car) one might check whether or not the pulsation is radially symmetrical.

Such a measurement of the distance of a Cepheid, if it can be carried out, is purely geometrical and independent of extinction and would represent a significant contribution to the calibration of their absolute magnitudes. It is also possible that the distance of fainter Cepheids might be found with reasonable accuracy by combining simple measurements of their average angular size with independent determinations of their radius by the Wesselink method (e.g. Fernie and Hube, 1967).

From this brief discussion we can see that the observations of Cepheid variables offer an astronomical programme of exceptional interest. The data might contribute to our understanding of how, and why, these stars pulsate and provide an independent calibration of their absolute magnitudes and hence of the astronomical distance scale.

12.5 *Emission-line Stars and Rotating Stars*

Our present knowledge of emission-line stars and of rapidly rotating stars depends heavily on theory and on the interpretation of the indirect evidence of photometry and spectroscopy. It seems that in some cases, for example Be stars, we cannot be sure what is taking place without a crude 'picture' of the star. Although a high-resolution optical interferometer cannot yet claim to give a 'picture', it certainly can yield some significant information about the main features of a star; more particularly it can be used to verify whether any particular theory about their structure is correct. For example, the work on γ Vel at Narrabri demonstrates how the size and crude shape of the emission region of a Wolf–Rayet star can be found. Admittedly it did not tell us with which of the two stars the emission region is associated, nevertheless it did provide a direct experimental check on the principal feature conventionally attributed to a Wolf–Rayet star.

In the study of stellar rotation one gains the impression that further decisive progress depends upon comparing the shape and brightness distributions of actual stars with the principal theories in the field. Single rotating stars present an attractive subject for study with an intensity interferometer, and it seems likely that a more powerful instrument could make significant contributions to this topic. Unfortunately we have been prevented from exploring this possibility satisfactorily at Narrabri because the work demands too high a signal/noise ratio for the present instrument. However, the application of high-resolution radio interferometry to the study of peculiar radio sources has proved a most fruitful line of work and this should encourage us to apply an optical interferometer to the study of peculiar stars.

12.6 Limb-darkening, Polarization and Extended Atmospheres

As we have already seen in §11.6, measurements of the limb-darkening of a single star can in principle be made with an intensity interferometer. However, this information is contained in the secondary maximum of the visibility curve and consequently, significant observations can only be made on stars which are four or five magnitudes brighter than the limit of the interferometer on single stars. Thus, an interferometer with a magnitude limit of $+7$ could be expected to measure the intensity distribution across single stars brighter than magnitude $+2$. This would include only about 50 stars, but it must be remembered that the existing evidence on limb-darkening is derived from eclipsing binaries and is far from adequate. It would therefore be valuable to investigate a few stars of different spectral types; in particular measurements of some giants and supergiants would be of great interest since virtually nothing is known about their limb-darkening.

Another interesting line of research with a large optical interferometer is suggested by the experiment with polarized light reported in §11.8. By measuring the apparent angular size of a star in two orthogonal polarizations, parallel and normal to the baseline, it should be possible to detect extended hot atmospheres by the effects of electron scattering. This question remains to be analysed in detail but, at first sight, there are some interesting possibilities. For example, a preliminary calculation suggests that one could measure the extended atmospheres around those OB supergiants which show spectroscopic evidence (Morton, 1967) of mass loss of the order 10^{-6} $M_\odot$ yr^{-1} at 1000 km s^{-1}.

12.7 Interstellar Extinction

One other possible application of a more sensitive interferometer, forseeable at present, would be the measurement of interstellar extinction. The aim of such a programme might be to measure the ratio of total to selective extinction, to establish whether there is a component of neutral extinction, and perhaps to study variations with direction over the galactic plane. To do this one might use a class of stars for which the surface flux F_λ is first standardized by measuring nearby stars. The angular sizes of more distant members of the same class would then be measured and, by combining these results with spectral scans, the apparent values of F_λ would be found for these distant stars and hence one could find the neutral and selective components of extinction. Ideally one would like to use luminous supergiants with absolute magnitudes as high as $-8\cdot0$; in the absence of interstellar extinction such stars could be measured at distances exceeding 10 kpc with an interferometer capable of reaching to $+7$ and, even with an extinction of $2\cdot0$ mag/kpc, one could reach distances up to 1·9 kpc. However, such stars are comparatively rare and a programme based on relatively common B 0 stars would allow measurements out to 2 kpc in the absence of extinction and to nearly 1 kpc for an absorption of 2·0 mag/kpc.

12.8 *The Specifications of a New Instrument*

The preceding review points firmly to the conclusion that any successor to the present intensity interferometer at Narrabri ought to have sufficient sensitivity and resolving power to measure single stars with a limiting magnitude of at least +7. Such an instrument would measure a star of magnitude +6 with a precision of ±5 per cent in two nights, a star of +5 in 2½ hours, and a star of +4 in about half an hour. To increase the limiting magnitude of the existing instrument from +2·5 to +7 (a factor of about ×60 in sensitivity) is not possible by any reasonable modification. We must therefore design a completely new instrument.

Two obvious ways of improving the sensitivity are to increase the reflector size and to increase the electrical bandwidth; it is not practicable to increase the exposure times compared with those at Narrabri because they are already inconveniently long. Another possible way is to increase the number of independent optical channels; in principle we can split the light from the star into a very large number of separate wavebands, each of which has its own phototubes and correlator. It was shown in §4.2.3 that the signal/noise ratio is independent of the optical bandwidth, provided that it is large compared with the electrical bandwidth; thus if we could split the light into m separate channels without loss, we should gain a factor of $(m)^{1/2}$ in the sensitivity.

Taking these three factors—area, bandwidth and number of optical channels—it is simple to show that the signal/noise ratio depends upon them as follows

$$S/N \propto A(b_v m)^{1/2} \tag{12.1}$$

where A is the geometric mean of the areas of the two reflectors, b_v is the overall electrical bandwidth of the correlator and the phototubes, and m is the number of independent optical channels. At Narrabri $A = 30$ m, $b_v \approx 60$ MHz, and $m = 1$.

Consider first the question of how much improvement we can expect to gain by increasing the electrical bandwidth. There appear to be two practical limits: the performance of available phototubes and the precision with which the path lengths and delays in the two arms of an interferometer can be maintained equal. The bandwidth of the present interferometer is limited by the phototubes to about 60 MHz but there are similar phototubes in existence with bandwidths of 300 MHz. Looking at more advanced phototubes, still in the experimental stage, it seems reasonable to assume that a bandwidth of about 1000 MHz or more will be possible in the not too distant future.

Turning to the limitations set by path differences and time delays, we saw in §4.2.4 that an increase in electrical bandwidth puts more stringent requirements on the precision of construction and movement of an interferometer. For example, if we make the electrical bandwidth 1000 MHz and require that any loss of correlation due to path differences or delays should be less than 1 per cent, then we must preserve the two paths through the instrument equal with a precision of ~ 1 cm, and any differential time delays in the electronics must be less than 0·3 ns. Experience at Narrabri suggests that these requirements might be met in a large instrument but that to demand an even higher precision is unrealistic and would endanger one of the main advantages of an intensity interferometer which is, of course, its freedom from the necessity of extremely high mechanical precision. Without more practical evidence it is not worth pursuing this topic further and we can only conclude, tentatively, that the bandwidth of an interferometer may be as high as 1000 MHz, giving an increase in sensitivity of $\times 4$ compared with the present instrument at Narrabri.

The possible number of independent optical channels presents an even more difficult question and can only be solved by detailed design and experiment. At first sight it lookes easy to split the light into several independent spectral bands illuminating separate detectors. However, a closer look shows that one cannot use conventional methods which depend upon angular dispersion, such as prisms and gratings, because of the imperfect collimation of the light from a large and necessarily crude reflector. It has been suggested that the problem might be solved by splitting the light into two orthogonally polarized beams and then passing each beam through a cascade of narrow-band interference filters; each filter would transmit its own band to an associated phototube and reflect the remainder of the light to the next filter. A preliminary analysis of this scheme suggests that ten independent optical channels might be obtained in this way. We shall therefore assume, again tentatively, that $m = 10$ and that the corresponding gain in sensitivity, relative to the Narrabri instrument, is $\times 3$.

Finally we arrive at the question, how large shall we make the reflectors? Clearly the answer will control to a large extent how much the instrument will cost and whether or not our whole programme is practical. Taking the proposed improvements in sensitivity of $\times 4$ due to electrical bandwidth and $\times 3$ due to several optical channels, we need to increase the area of the reflectors by a factor of $\times 5$ ($A = 150$ m^2) to reach the required overall improvement of $\times 60$ in the sensitivity. Are such large reflectors feasible? An effective discussion of this point needs to be focused on to some specific configuration.

12.9   A Possible Configuration of a New Interferometer

Any proposed configuration of a new instrument capable of resolving single stars of magnitude $+7$ must have, in addition to the three major

N

requirements just discussed in §12.8, a maximum baseline of at least 2 km; to exclude excessive background light from the night-sky the angular diameter of the field of view of each reflector must be limited to about 3 minutes of arc; to avoid loss of correlation path differences between the two arms of the interferometer must be less than ∼ 1 cm.

Fig. 12.2 illustrates one-half of the configuration which has been proposed by the Chatterton Astronomy Department of the University of Sydney. In fig. 12.2 light from the star is received on the flat mirrors or coelostats (F_1, F_2) and is reflected to the fixed paraboloidal reflectors (R_1, R_2) which focus it on to the fixed detectors (D_1, D_2). The coelostats are mounted on trucks which roll on straight rails running east and west. As the star moves over the sky the two coelostats move so that the *projected baseline* (AB in fig. 12.2) *remains constant* and *the light from the star arrives at the two detectors at the same time*, so that

$$AF_1 + F_1R_1 = BF_2 + F_2R_2. \tag{12.2}$$

This basic configuration has a number of advantages; the most important are that the projected baseline can be held constant throughout an observation; there is no relative time delay in the light reaching the two detectors and the complexity of a wideband multi-channel

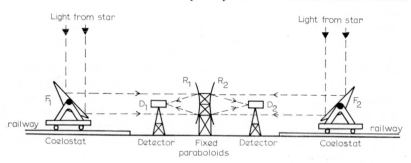

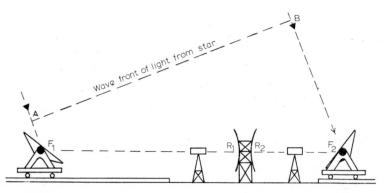

Fig. 12.2. A proposed configuration for a large intensity interferometer.

170

variable electrical delay is avoided; the optical systems are fixed and accessible and can have short, fixed leads to the correlator; the baselines can be extended at any time by laying more railway track. It has the disadvantages that the rail track is necessarily longer than the projected baseline; the light from the star must travel along the baseline thereby increasing atmospheric extinction and scintillation and, for very long baselines, introducing some differential velocity dispersion. Nevertheless, this proposed configuration seems preferable to all the alternatives which one can visualize. The obvious alternative—the circular layout which has worked so well at Narrabri—is not attractive for a larger instrument on several grounds, the most important objection being that the maximum resolving power is permanently fixed by the diameter of the circular track and cannot be increased without a major reconstruction.

As it stands, the configuration shown in fig. 12.2 would not be satisfactory for the very large reflectors (150 m²) that we propose. The individual reflectors and coelostats would be inconveniently large; their size would seriously complicate the design of the optical system at the focus; and they would partially resolve many of the stars and thereby complicate the interpretation of the measurements. For this reason it is preferable to make each unit half the required size and to use a system of four reflectors and coelostats, each pair running on parallel tracks.

The proposed coelostats and reflectors are circular in outline and, after allowance is made for the reduction in their effective area due to the inclination of the coelostats, it is found that they must have a diameter of about 12 m. From existing experience of building large mosaic reflectors, for example solar furnaces, one can say that the proposed fixed reflectors can be built without much difficulty and at a reasonable cost. The critical question is whether or not the moving coelostats can also be built at a reasonable cost and with sufficient surface accuracy to permit the use of a 3 minutes of arc angular field of view. In an attempt to answer this vital point a preliminary structural design of the coelostats was carried out which showed that the required performance can readily be achieved by a steel structure of very moderate mass (about 16 Mg) supporting polished aluminium panels.

Many other problems of detail such as the precision of the rail track, positioning and control of the coelostats, the design of a multi-channel correlator and the design of a multi-channel optical system have also been studied. All our investigations lead us to conclude that the instrument could be built, and that the overall cost would be reasonable —roughly that of a 2 m telescope complete with dome and accessories.

12.10 *Summary*

In this chapter we have argued that, following the pioneer work of the stellar interferometer at Narrabri, a high-resolution interferometer is an important tool in stellar astronomy.   Some of the most interesting applications (e.g. the study of Cepheid variables) remain unexplored. To carry on this work requires a more sensitive successor to the Narrabri interferometer and we have suggested that the next stage in this work should be to build a new instrument which will reach to stars of magnitude $+7$.   Preliminary consideration has been given to the design and cost of such an instrument and it has been concluded, tentatively, that it could be built, and that it would cost roughly as much as a conventional 2 m telescope.

At the present time this proposed instrument is no more than some drawings and a model.   Nevertheless, the scientific rewards of carrying on the work of the Narrabri stellar interferometer would be great; one can only hope that in due course the opportunity to do so will arise.

REFERENCES

ÁDÁM, A., JÁNOSSY, L. and VARGA, P., 1955, Observations on coherent light beams by means of photomultipliers, *Acta phys. Hung.*, **4**, 301.

ALLEN, C. W., 1963, *Astrophysical Quantities*, second edition, Athlone Press, London.

ALLEN, L. R. and FRATER, R. H., 1970, Wideband multiplier correlator, *Proc. Instn elect. Engrs*, **117**, 1603.

BERAN, M. J. and PARRENT, G. B., 1964, *Theory of Partial Coherence*, Prentice-Hall, Englewood Cliffs, N.J.

BORN, M. and WOLF, E., 1959, *Principles of Optics*, Pergamon Press, London.

BRANNEN, E. and FERGUSON, H. I. S., 1956, Photon correlation in coherent light beams, *Nature*, **178**, 481.

BRANNEN, E., FERGUSON, H. I. S., and WEHLAU, W., 1958, Photon correlation in coherent light beams, *Can. J. Phys.*, **36**, 871.

BRAMLEY, E. N., 1955, Some aspects of the rapid directional fluctuations of short radio-waves reflected at the ionosphere, *Proc. Instn elect. Engrs* B, **102**, 533.

CHANDRASEKHAR, S., 1946, On the radiative equilibrium of a stellar atmosphere, *Astrophys. J.*, **103**, 351.

CONTI, P. S. and SMITH, L. F. 1972, The absolute magnitudes and spectral types of the stars in the gamma Velorum system, *Astrophys. J.*, **172**, 623.

DAVIS, J., MORTON, D. C., ALLEN, L. R. and HANBURY BROWN, R., 1970, The angular diameter and effective temperature of zeta Puppis, *Mon. Not. R. astr. Soc.*, **150**, 45.

EINSTEIN, A., 1915, Antwort auf eine Abhandlung M. von Laues, "Ein Satz der Wahrscheinlichkeitsrechnung und seine Anwedung auf die Strahlungstheorie, *Ann. Phys.*, **47**, 853.

EINSTEIN, A. and HOPF, L., 1910, Uber einen Satz der Wahrscheinlichkeitsrechnung und seine Anwedung in der Strahlungstheorie, *Ann. Phys.*, **33**, 1096.

FARKAS, G. L., JÁNOSSY, L., NÁRAY, Z. and VARGA, P., 1965, Intensity correlation of coherent light beams, *Acta phys. Acad. scient. Hung.*, **18**, 199.

FELLGETT, P., 1957, The question of correlation between photons in coherent beams of light, *Nature*, **179**, 956.

FELLGETT, P., CLARK JONES, R. and TWISS, R. Q., 1959, Fluctuations in photon streams, *Nature*, **184**, 967.

FERNIE, J. D. and HUBE, J. O., 1967, On Wesselink's method of determining stellar radii, *Publ. astr. Soc. Pacific*, **79**, 95.

FORRESTER, A. T., GUDMUNDSEN, R. A. and JOHNSON, P. O., 1955, Photoelectric mixing of incoherent light, *Phys. Rev.*, **99**, 1691.

FRANÇON, M., 1966, *Optical Interferometry*, Academic Press, New York.

FRATER, R. H., 1964, Accurate wideband multiplier and square-law detector, *Rev. scient. Instrum.*, **35**, 810; 1965, Synchronous integrator and demodulator, *Ibid.*, **36**, 634.

GINGERICH, O., 1969, *Theory and Observation of Normal Stellar Atmospheres*, M.I.T. Press, Cambridge, Mass.

HANBURY BROWN, R. and BROWNE, A., 1966, The stellar interferometer at Narrabri Observatory, *Philips tech. Rev.*, **27**, 141.

HANBURY BROWN, R., DAVIS, J. and ALLEN, L. R., 1967, The stellar interferometer at Narrabri Observatory—I, *Mon. Not. R. astr. Soc.*, **137**, 375; 1969, The effect of Čerenkov light pulses on a stellar intensity interferometer, *Ibid.*, **146**, 399.

HANBURY BROWN, R., DAVIS, J. and ALLEN, L. R., 1974, The angular diameters of 32 stars, *Mon. Not. R. astr. Soc.*, **167**, 121.

HANBURY BROWN, R., DAVIS, J., ALLEN, L. R., and ROME, J. M., 1967, The stellar interferometer at Narrabri Observatory—II, *Mon. Not. R. astr. Soc.*, **137**, 393.

HANBURY BROWN, R., DAVIS, J., HERBISON-EVANS, D. and ALLEN, L. R., 1970, A study of γ^2 Velorum with an intensity interferometer, *Mon. Not. R. astr. Soc.*, **148**, 103.

HANBURY BROWN, R., HAZARD, C., DAVIS, J. and ALLEN, L. R., 1964, A preliminary measurement of the angular diameter of α Lyrae, *Nature*, **201**, 1111.

HANBURY BROWN, R., JENNISON, R. C. and DAS GUPTA, M. K., 1952, Apparent angular sizes of discrete radio sources, *Nature*, **170**, 1061.

HANBURY BROWN, R. and TWISS, R. Q., 1954, A new type of interferometer for use in radio-astronomy, *Phil. Mag.*, **45**, 663; 1956 a, Correlation between photons in two coherent beams of light, *Nature*, **177**, 27; 1956 b, A test of a new type of stellar interferometer on Sirius, *Ibid.*, **178**, 1046; 1956 c, The question of correlation between photons in coherent light rays, *Ibid.*, **178**, 1447; 1957 a, The question of correlation between photons in coherent beams of light, *Ibid.*, **179**, 1128; 1957 b, Interferometry of the intensity fluctuations in light. Part I. Basic theory: the correlation between photons in coherent beams of radiation, *Proc. R. Soc.* A, **242**, 300; 1957 c, Part II. An experimental test of the theory for partially coherent light, *Ibid.*, **243**, 291; 1958 a, Part III. Applications to astronomy, *Ibid.*, **243**, 199; 1958 b, Part IV. A test of an intensity interferometer on Sirius A, *Ibid.*, **248**, 222.

HARDORP, J. and SCHOLZ, M., 1971, Effect of rapid rotation on radiation from stars, *Astr. Astrophys.*, **13**, 353.

HARDORP, J. and STRITTMATTER, P. A., 1968, Effect of rapid rotation on radiation from stars, *Astrophys. J.*, **153**, 465.

HARRINGTON, J. P., 1970, Polarization of radiation from stellar atmospheres. The grey case. *Astrophys. Space Sci.*, **8**, 227.

HARRIS, D. L., 1963, *The Stellar Temperature Scale and Bolometric Corrections*, in *Basic Astronomical Data*, Ed. K. Aa. Strand, University of Chicago Press, Chicago, p. 263.

HARRIS, D. L., STRAND, K. AA., WORLEY, C. E., 1963, *Empirical Data on Stellar Masses, Luminosities and Radii, in Basic Astronomical Data*, Ed. K. Aa. Strand, University of Chicago Press, p. 273.

HEINTZE, J. R. W., 1968, Temperature, gravity and mass of Vega, Sirius and τ Herculis, *Bull. astr. Inst. Netherl.*, **20**, 1.

HERBISON-EVANS, D., HANBURY BROWN, R., DAVIS, J. and ALLEN, L. R., 1971, A study of α Virginis with an intensity interferometer, *Mon. Not. R. astr. Soc.*, **151**, 161.

HILL, D. A. and PORTER, N. A., 1961, Photography of Čerenkov light from extensive showers in the atmosphere, *Nature*, **191**, 690.

HODARA, H., 1966, Laser Wave Propagation through the atmosphere, *Proc. Instn elect. Engrs*, **54**, 368.

HOFFLEIT, D., 1964, *Catalogue of Bright Stars*, Yale University Observatory.

IRELAND, J. G., 1966, The effect of rotation on stellar luminosity, *Pub. R. Obs. Edinb.*, **5**, 63.

JELLEY, J. V. and GALBRAITH, W., 1955, Light pulses from the night sky and Čerenkov radiation, *J. atmos. terr. Phys.*, **6**, 304.

JENKINS, L. F., 1963, *Supplement to the General Catalogue of Trigonometric Stellar Parallaxes*, Yale University Observatory.

JENNISON, R. C. and DAS GUPTA, M. K., 1956, The measurement of the angular diameters of two intense radio sources, *Phil. Mag.* Ser. (8), **1**, 66.

JOHNSTON, I. D. and WAREING, N. C., 1970, On the possibility of observing interferometrically the surface distortion of rapidly rotating stars, *Mon. Not. R. astr. Soc.*, **147**, 47.

LAMB, W. E. and SCULLY, M. O., 1969, *The photoelectric effect without photons*, in *Polarisation, Matière et Rayonnement*, Presses Universitaires de France, Paris, p. 363.

LAUE, M. VON, 1915 a, Ein Satz der Wahrscheinlichkeitsrechnung und seine Anwedung auf die Strahlungstheorie, *Ann. Phys.*, **47**, 879; 1915 b, Zur Statistik der Fourier Koeffizienten der naturlichen Strahlung, *Ibid.*, **48**, 668.

LINDENBLAD, I. W., 1970, Relative photographic positions and magnitude difference of the components of Sirius, *Astr. J.*, **75**, No. 7, 841.

MANDEL, L., 1963, Fluctuations of light beams, *Progress in Optics*, Ed. E. Wolf, North-Holland, Amsterdam, p. 181.

MANDEL, L., SUDARSHAN, E. C. G., and WOLF, E., 1964, Theory of photoelectric light fluctuations, *Proc. phys. Soc.*, **84**, 435.

MARTIENSSEN, W. and SPILLER, E., 1964, Coherence and fluctuations in light beams, *Am. J. Phys.*, **32**, 919.

MICHELSON, A. A. and PEASE, F. G., 1921, Measurement of the diameter of α Orionis with the interferometer, *Astrophys. J.*, **53**, 249.

MIKESELL, A. H., HOAG, A. A. and HALL, J. S., 1951, Scintillation of starlight, *J. opt. Soc. Am.*, **41**, 689.

MORRIS, G. J., 1971, Observations of intensity modulation of starlight at discrete radio frequencies, *Phys. Rev. Lett.*, **27**, 1600.

MORTON, D. C., 1967, Mass loss from three OB supergiants in Orion, *Astrophys. J.*, **150**, 535.

NELDER, J. A. and MEAD, R., 1965, A simplex method for function minimization, *Comput. J.*, **7**, 308.

PEASE, F. G., 1925, The fifty-foot interferometer telescope of the Mount Wilson Observatory, *Armour Engr*, **16**, 125; 1930, The new fifty-foot stellar interferometer, *Scient. Am.*, **143**, 290; 1931, Interferometer methods in astronomy, *Ergebn. Exacten Naturwiss.*, **10**, 84.

PURCELL, E. M., 1956, The question of correlation between photons in coherent light rays, *Nature*, **178**, 1449.

175

REBKA, G. A. and POUND, R. V., 1957, Time-correlated photons, *Nature*, **180**, 1035.

RICE, S. O., 1944, Mathematical analysis of random noise, *Bell Syst. tech. J.*, **23**, 1, and **24**, 46.

SMAK, J., 1967, The spectroscopic orbit of CO Lacertae, *Acta astr.*, **17**, 245.

SMITH, L. F., 1968, *The features of the system of Wolf–Rayet stars, in Wolf–Rayet stars*, Ed. K. B. Gebbie and R. N. Thomas, National Bureau of Standards Special Publication 307, Washington, D.C., p. 23.

STEEL, W. H., 1967, *Interferometry*, Cambridge University Press.

TWISS, R. Q. and LITTLE, A. G., 1959, The detection of time-correlated photons by a coincidence-counter, *Austr. J. Phys.*, **12**, 77.

TWISS, R. Q., LITTLE, A. G. and HANBURY BROWN, R., 1957, Correlation between photons in coherent light rays, *Nature, Lond.*, **178**, 1447.

WEBB, R. J., 1971, Angular diameters and fluxes of early-type stars. Ph.D. Thesis, University of Sydney.

YERBURY, M. J., 1967, Amplitude limiting applied to a sensitive correlation detector, *Radio electron. Engr*, **34**, 5; 1968, Experimental tests of the theory of an amplitude-limited correlation detector, *Ibid.*, **36**, 71.

ZATSEPIN, V. I., 1964, The angular distribution of intensity of Čerenkov radiation from extensive air showers, *Zh. eksp. teor. Fiz.*, **47**, 689.

ABBREVIATIONS

of constellations used in the text

Aql	Aquila	(Eagle)
Boo	Boötes	(Herdsman)
CMa	Canis Major	(Great dog)
CMi	Canis Minor	(Small dog)
Car	Carina	(Keel)
Cen	Centaurus	(Centaur)
Cet	Cetus	(Whale)
Cru	Crux	(Southern Cross)
Crv	Corvus	(Crow)
Eri	Eridanus	(River Eridanus)
Gem	Gemini	(Heavenly twins)
Gru	Grus	(Crane)
Leo	Leo	(Lion)
Lyr	Lyra	(Lyre)
Oph	Ophiucus	(Serpent bearer)
Ori	Orion	(Hunter)
Pav	Pavo	(Peacock)
Peg	Pegasus	(Winged Horse)
PsA	Piscis Austrinus	(Southern fish)
Pup	Puppis	(Poop, stern)
Sgr	Sagittarius	(Archer)
Sco	Scorpius	(Scorpion)
Tau	Taurus	(Bull)
Vel	Vela	(Sails)
Vir	Virgo	(Virgin)

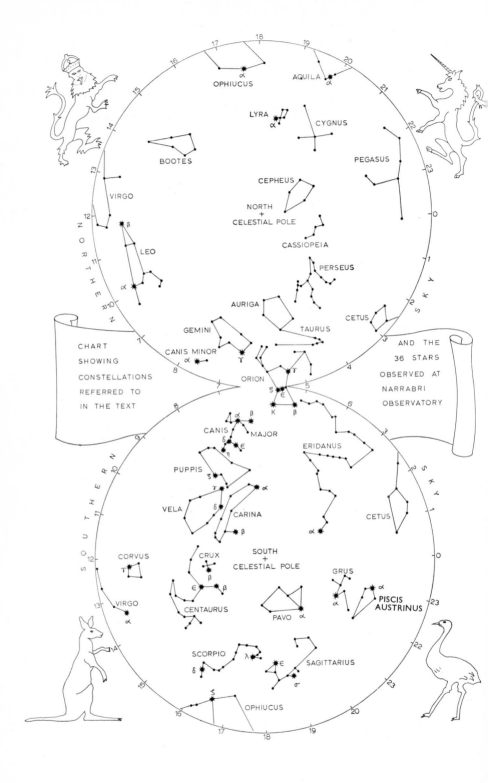

CHART SHOWING CONSTELLATIONS REFERRED TO IN THE TEXT AND THE 36 STARS OBSERVED AT NARRABRI OBSERVATORY

THE NAMES OF STARS
referred to in the text (from Allen, 1963)

α Aql,	Altair		α Lyr,	Vega
η Aql,	——		α Oph,	Ras-Alhague
α Boo,	Arcturus		ζ Oph,	Alnitak
α CMa,	Sirius		α Ori,	Betelgeuse
β Cma,	Mirzam		β Ori,	Rigel
δ CMa,	Wezen		γ Ori,	Bellatrix
ε CMa,	Adhara		ζ Ori,	Alnitak
η CMa,	Aludra		ε Ori,	Alnilam
α CMi,	Procyon		κ Ori,	Saiph
α Car,	Canopus		ξ Ori,	—
β Car,	Miaplacidus		α Pav,	Peacock
ρ Car,	—		β Peg,	Scheat
β Cen,	Hadar		α PsA,	Fomalhaut
ε Cen,	—		ζ Pup,	Naos
σ Cet,	—		σ Sgr,	Nunki
γ Cru,	Gienah		ε Sgr,	Kaus-Australis
β Cru,	Mimosa		α Sco,	Antares
α Eri,	Achernar		δ Sco,	Dzuba
γ Gem,	Alhena		λ Sco,	Shaula
ζ Gem,	—		α Tau,	Aldebaran
α Gru,	Al Na'ir		δ Vel,	—
α Leo,	Regulus		γ Vel,	—
β Leo,	Denebola		α Vir,	Spica

179

INDEX

Amplifiers, 76–77, 103–104
Analytic signal, 32
Angular size
equivalent uniform disc, 114, 122–123, 125, 132–135, Table 11.1
limb-darkened star, 123, 125, fig. 10.4, 134–135, Table 11.1
radio sources, 2, 3, 84–88
Atmospheric extinction, 16, 90, 116, 132, 160–161
Atmospheric irregularities, 68–69, 71–72
Atmospheric scintillation
amplitude, 70–72
angular, 71–72
correlation length, 72
Michelson's interferometer, 2, 25, 43–45, 51, 162
radio intensity interferometer, 4, 86–88
optical intensity interferometer, 4, 9, 18, 52, 54, 67–72, 87, 120, 159–160, fig. 11.9
optical telescope, 2
phase, 68–69
Australian Research Grants Committee, vii
Auto-correlation function, 38, 54

Bandwidth
electrical, see Electrical
optical, see Light
Balmer jump, 138
Baseline
choice, 114–115, fig. 9.2, 127–128
misalignment, 121–122
for proposed new instrument, 170
Bessel function, 43
Binary stars, 18, 20, 27, 126–128, 132–133, 141–148, 163–165
Bose–Einstein statistics, 6, 31, 55

Catenary cable, 19, 94
Cepheid variable, 145, 165–166, 172
Čerenkov light, 18, 64–67, 121, 154–159
Chatterton Astronomy Department (Sydney), 170

Coherence
complex degree of coherence, 27–29, 34–37, 42–44, 50–52
degree of coherence, 24, 26, 40, 42, 57, 60–61
length, 36, 39, 49
mutual coherence function, 34, 36, 38
partial coherence factor, 60–63, fig. 5.1, 123
spatial coherence, 35–37, 53
temporal coherence, 35–38, 44
time, 39, 55, 57, 81.
Coincidence
counter, 54–58
counting, 10, 56–58, 79–83
counting interferometer, 57, fig. 4.9
excess, 58, 81–83
random, 81–83
Computer, see Control
Computer programme, 142
Control
building, 94, 100
computer, 100, 116, 122
desk, 94, fig. 8.7, 115–116
of reflectors, 100–101
Correlation
and aperture size, 59–62
and baseline length, 27, fig. 2.3, 50, 87, fig. 7.3, 124–128, fig. 10.2
and Čerenkov light, 65–66, fig. 5.2
and degree of coherence, 29
and polarization, 40, 57
and radio interferometer, 86
correction for zero-drift, 118
cumulative, 116–117, fig. 9.3
factor, 60, 78–79, fig. 6.3
false or spurious, 103, 104, 107–108, 121, fig. 10.1, 157–158
laboratory measurement, 8, 9, 10
normalized, 79, 86–87, 117–119, 122, 142
of intensity fluctuations in light, 3, 39–40
of fluctuations in photoelectric detectors, 48–50
uncertainty, 113–114, 118–121
zero baseline, 92, 122–123, 127–128, 132–135, Table 11.1, 140, Table 11.4

Correlator
 amplitude-limiting, 109–110
 arrival at Narrabri, 16
 bandwidth, 60, 63
 contract to develop, 12
 dummy runs, 74, 113, 117, fig. 9.3
 efficiency, 161
 excess noise, 160
 gain, 118–120
 reliability, 16
 stability, 16–17, 108, 110
 used in early laboratory experiments, 74–75, fig. 6.2
 used in Narrabri interferometer,
  103–109, fig. 8.8
 zero drift, 74–75, 104, 107–108,
  fig. 8.11, 118–119
Cosmic rays, 64, 155–157

Data-handling, 106–107
Discriminator, 155
Dispersion of light, 53, 69, 70
D.S.I.R., vi, 11, 12
Double stars, see Binary stars
Dunford and Elliott, vi, 11, 12

Effective temperature, see Stellar
Electrical
 bandwidth, 3, 46–47, 50, fig. 4.4,
  60, 70, 168–169
 delay, 81, 112–113, fig. 9.1, 171
 fluctuations, 26–29, 45–50
 spectral density of fluctuations, 52,
  60, 63, 160
Electron-scattering, 129–131, fig. 10.5,
 167
Emergent flux, see Stellar
Emission-line stars, 18, 148–149
Exposure time, 114–115, fig. 9.2
Extended atmospheres, 167

Faraday rotation, 87
Fluctuations
 intensity of light, 26, 39, 40
 photoelectric currents, 26, 45–47
 temperature of grey body, 6
 photon-counting, 54–55

Flux, see Stellar
Foundation for Research in Physics,
 vi
Fourier
 analysis, 28, 68
 components, 28–32, 46, 123
 integral, 32–33
 spectroscopy, 38
 transform, 37, 38, 42, 52, 61, 123,
  125, 129
Fringe
 formation, 22–25, fig. 2.2, 41–42,
  fig. 4.1
 visibility, 22–25, 29, fig. 2.3, 41–45,
  fig. 4.2

Gaussian noise, 30, 32, 39
Guiding, see Star-guiding

Harvard, 10

Image size, see Reflectors
Integrating motor, 75–77, 90
Integrators,
 current, 103, 107, 120
 synchronous, 106, fig. 8.10
Intensity interferometer
 advantages, 4, 29, 30
 coincidence-counting, 57, fig. 4.9,
  79–83, fig. 6.4
 disadvantages, 51
 effect of atmospheric scintillation,
  see Atmospheric scintillation
 effect of path differences or time
  delays, 29–30, 51–53, fig. 4.7,
  69–70
 pilot model, 88–90
 principles, 25–31, fig. 2.5, 45–50,
  fig. 4.5
 proposed new instrument, 168–172
 radio, 2, 3, 84–88, fig. 7.1
 using linear multiplier, 25–31, figs.
  2.4, 2.5, 45–50, fig. 4.5, 73–79
Interference filter, 28, 88, 89, 98–99,
 fig. 8.4, 148–149, 169
Interstellar extinction, 167
Ionosphere, 87, 88, 157
Isotope lamp, 80, 82

181

Jodrell Bank, 3, 8, 9, 10, 11, 12, 13, 14, 19, 73, 89, 90

Lasers, 69
Light
  bandwidth, 28, 44, 45, 49, 50, 58, 59, 90, 98, 120, 155
  coherence, *see* Coherence
  coherence length, 36, 39, 49
  coherence time, 55
  mathematical description, 32–40
  mutual spectral density, 38, 39, 52
  spectral density, 33, 55, 60
  wave-particle duality, 83
Limb darkening, *see* Stellar
Lindars Automation, 12
Linear multiplier, *see* Multiplier

MacDonald, Wagner and Priddle, vii, 14
Mass loss from stars, 131, 154–155, 167
Michelson's interferometer
  at Mt. Wilson, 2, 24
  effect of atmospheric scintillation, *see* Atmospheric scintillation
  effect of path differences or time delays, 24–25, 43–45, fig. 4.2, 51, 53
  fifty ft. model at Mt. Wilson, 2
  future developments, 2, 162
  principles, 22–25, 41–45, fig. 4.1
  results, 2, 163, fig. 12.1
Mirrors, *see* Reflectors
Mount Wilson, 2, 24, 25
Mullards Ltd., vii, 12, 16
Multiplier, 16, 17, 45–48, 58, 73–74, 85, 104–105, fig. 8.9, 107
Multiple stars, 139–141, also *see* Binary stars
Mutual spectral density, 38, 52

Narrabri stellar interferometer
  construction, 93
  correlator, 103–109, fig. 8.8
  cost, 11, 17
  general layout, 94–95, fig. 8.1
  installation, 14–16
  parameters, 160

pilot programme, 16–17
preliminary test, 16
results of main programme, 20, 132–161, Table 11.1
sensitivity, 16, 18
signal-to-noise, 160–161, fig. 11.10
site, 14, 94–95

Noise
  excess noise in phototubes, *see* Phototubes
  in coincidence counting, 54–58
  in output of correlator, 50, 63, 76
  in output of phototube, 45–48
  shot noise, 25–26, 47–48, 55, 74, 86
  signal-to-noise ratio, *see* Signal-to-noise ratio
  uncertainty due to noise, 63, 113–114
  wave noise, 26–27, 47–48, 55, 74, 86
Nuclear Research Foundation, vi

Office of Scientific Research (U.S. Air Force), vii, 17
Officine Galileo, 12, 13
Optical bandwidth, *see* Light
Optical filters, *see* Interference filters
Optical system
  coincidence-counting experiment, 80, fig. 6.4
  early test of correlation, 74, fig. 6.1
  Narrabri stellar interferometer, 98–99, fig. 8.3
  pilot model stellar interferometer, 88, figs. 7.4, 7.5
  proposed new instrument, 169

Parallax, *see* Stellar
Paschen continuum, 138
Path difference, *see* Time delay, Michelson's interferometer, Intensity interferometer
Phase-switching, 74–77, 103, 106–107, 158
Photoelectric emission
  probability, 54–56, fig. 4.8, 68, 70
  semi-classical model, 5, 45–46, 49, 88

182

Photocathode
  quantum efficiency, 50, 54, 74, 90,
    102, 103
  size, 99, 102, 112
  spectral response, 74, 90
Photon
  arguments about, 5–9
  bunching, 7, 31, 55
  Čerenkov, 65–65
  correlation and coincidences, 7, 8,
    9, 79–83
  radio, 4, 5
Phototube
  advances, 168
  excess noise, 55, 63, 102, 160
  response time and bandwidth, 49,
    63, 102
  types, 74, 90, 102
Photomultiplier, see Phototube
Polarization, 18, 20, 57, 60, 129–130,
  fig. 10.5, 154–155, 160, 167
Position angle
  α CMa, 151
  α Aql, 151–154, fig. 11.6
  α Vir, 141–145, Table 11.5
  binary stars, 126–127

Quantum efficiency, see Photocathode
Quantum theory, 8, 9, 79–80

Radio interferometer, see Intensity
  interferometer
Radio-frequency modulation of light,
  157–158
R.C.A., 17, 74, 90, 102
Radio sources, 3–4, 84, 86, fig. 7.3
Radio interference, 151, 157–158
Random coincidences, see Coincidence
Reflectors
  alignment, 15, 111–112
  cleaning, 96
  cost, 12
  damage to mirrors, 15
  effect of dew, 10, 89, 97
  field of view, 66, 67, 112, 155, 171
  first test, 15, 111
  image size, 5, 15, 89, 99, 102, 111–
    112
  mirror construction, 13, 96
  mirror specification, 15, 96

Narrabri interferometer, 94–96,
  figs. 8.2, 8.5, 8.6
pilot interferometer, 89–90, 111,
  fig. 7.5
proposed for large interferometer,
  171
searchlights, 89–90, 111, fig 7.5
Refractive index, 53, 69
Rotation of stars, see Stellar
School of Physics (University of
  Sydney), 11, 17
Scintillation, see Atmospheric scintillation
Shot noise, see Noise
Signal-to-noise
  amplitude-limiting correlator, 109–
    110
  maximum possible, 63–64
  Narrabri interferometer see Narrabri
    stellar interferometer
  optical intensity interferometer, 10,
    20, 50–51, 58, 63–64, 78, 91–92,
    102, 159–161, fig. 11.10
  radio interferometer, 86
Spectroscopic binary, 141–148, 164–
  165
Star
  α Aql, Altair, 129, 135, 136, 137,
    151, 152, 153
  η Aql, —, 166
  α Boo, Arcturus, 25
  α CMa, Sirius, 9, 88, 90, 91, 92, 93,
    103, 122, 123, 134, 136, 137, 149,
    150, 151, 159
  β CMa, Mirzam, 134
  δ CMa, Wezen, 18, 134
  ε CMa, Adhara, 134
  η CMa, Aludra, 134, 161
  α CMi, Procyon, 134, 136, 137
  α Car, Canopus, 103, 122, 123, 133,
    134, 136, 137, 149
  β Car, Miaplacidus, 134, 136
  ρ Car, —, 166
  β Cen, Hadar, 16, 141, 158
  ε Cen, —, 135
  σ Cet, —, 25
  γ Crv, Gienah, 135, 161
  β Cru, Mimosa, 18, 116, 124, 133,
    135, 140, 158

α Eri, Achernar, 110, 134
γ Gem, Alhena, 134, 136
ζ Gem, —, 166
α Gru, Al Na'ir, 124, 135, 136
α Leo, Regulus, 129, 134, 136
β Leo, Denebola, 134, 136
α Lyr, Vega, 1, 2, 16, 102, 124, 132, 135, 136, 138, 158
α Oph, Ras-Alhague, 25, 135, 136, 161
ζ Oph, Altnitak, 135, 161
α Ori, Betelgeuse, 25
β Ori, Rigel, 131, 134, 137, 138, 154
γ Ori, Bellatrix, 134
ζ Ori, Alnitak, 133, 134, 140
ε Ori, Alnilam, 134
κ Ori, Saiph, 134
ξ Ori, —, 133
α Pav, Peacock, 135
β Peg, Scheat, 25
α PsA, Fomalhaut, 135, 136
ζ Pup, Naos, 134, 136, 163
σ Sgr, Nunki, 128, 141
ε Sgr, Kaus-Australis, 135
α Sco, Antares, 25
δ Sco, Dzuba, 127, 133, 135, 140
λ Sco, Shaula, 128, 141
α Tau, Aldebaran, 25
δ Vel, —, 141
γ Vel, —, 133, 134, 136, 140, 141, 148, 166
α Vir, Spica, 20, 128, 133, 135, 136, 140, 141, 145, 146, 148
Star-guiding, 15, 100–102, 112, 116
Stellar
  absolute magnitude, 142, 147–148, fig. 11.4, 166
  angular diameter, *see* Angular size
  corona, 131, 154–155
  effective temperature, Table 11.1, 137–139, 163–164
  emergent flux, 136, 163
  emission lines, 18, 148–149
  limb-darkening, 18, 123–126, figs. 10.3, 10.4, 129–130, 145, 149–151, 167
  luminosity, 134–136, 145, 147, 163–164
  mass, 142, 145, 147

mass–luminosity relation, 147–148, fig. 11.4
model atmosphere, 136–139, fig. 11.1
parallax (distance), 133, 136, 145, 146–148, 165–167
polarization, *see* Polarization
radius, 133, 136, Table 11.2, 145, 147, 163–164
rotation, 18, 128–129, 145, 151–153, 166
spectral type, 132, 134, 135, Table 11.1, 164
surface gravity, 128, 133, 138, 145, 147
Supergiants, 131, 133, 154, 167
Synchronous integrator, *see* Integrator
Synchronous rectifier (demodulator), 74–76, 106

Temperature, *see* Stellar
Time delay, 44, 52–53, fig. 4.7, 69, 90, 112, fig. 9.1, 121, *see* Michelson's interferometer, Intensity interferometer
Track, 14, 15, 102, 170–171, 94–95, fig. 8.1

University
  California, 17
  Harvard, 10
  Manchester, 11, 17, 18, 73, 89
  Michigan, vii
  Princeton, vii
  Sydney, 11, 14, 17, 18
  Western Ontario, 8

van Cittert—Zernike theorem, 35, fig. 3.2, 37
von Zeipel's theorem, 128

Wave noise, 5, 6
Wesselink's method, 166
Wiener-Khinchin theorem, 38, 52
Wind-speed, 116, 117
Wolf-Rayet star, 166

Zero-drift in correlator, 74–75, 104, 107–108, fig. 8.11, 109–110, 112–113, 118–119
Zero-level uncertainty, 121, fig. 10.1